A. Friedberg Stanton

Hog Cholera

Its history, nature and treatment, as determined by the inquiries and investigations

of the Bureau of animal industry

A. Friedberg Stanton

Hog Cholera
Its history, nature and treatment, as determined by the inquiries and investigations of the Bureau of animal industry

ISBN/EAN: 9783337241339

Printed in Europe, USA, Canada, Australia, Japan

Cover: Foto ©berggeist007 / pixelio.de

More available books at **www.hansebooks.com**

HOG CHOLERA:

ITS HISTORY, NATURE, AND TREATMENT,

AS DETERMINED BY

THE INQUIRIES AND INVESTIGATIONS

OF

THE BUREAU OF ANIMAL INDUSTRY.

———◆•◆•◆———

WASHINGTON:
GOVERNMENT PRINTING OFFICE.
1889.

15612 π c——1

TABLE OF CONTENTS.

4

ILLUSTRATIONS.

5

LETTER OF TRANSMITTAL.

WASHINGTON, D. C., *January* 18, 1889.

SIR : I have the honor to submit herewith a report upon the history nature, and treatment of the disease known in the United States as hog cholera. Our knowledge of this pest has been developed almost entirely by the inquiries and experimental investigations of the Bureau of Animal Industry; and while much of the information contained in this volume has been published in the reports of the Department of Agriculture, a systematic and complete statement of the facts has never before been made in a connected manner.

It has been discovered in the course of these investigations that there are two very different and distinct epizootic diseases of swine in this country which are widely prevalent, and which had previously been spoken of under the one name of hog cholera or swine plague. These two names had, therefore, been used synonymously previous to 1886, when the differences between the diseases were pointed out in the reports of this Bureau. It was then deemed best to apply the term hog cholera to that disease in which the intestines were found most affected, and in which the symptoms would come nearest to justifying the appellation. The other malady appeared identical, not only in symptoms and lesions but in the microbe which caused it, with the disease recently described in Germany by both Löffler and Schütz under the name of *Schweineseuche*, or swine plague. For this reason it was considered best to call this affection swine plague in the reports, and thus preserve uniformity and prevent confusion of ideas in reference to the diseases of swine in different countries.

This volume treats exclusively of hog cholera, as it is the malady which has been longest under investigation; but the material is on hand for an almost equally complete volume on swine plague, which we hope soon to submit for publication. There are, of course, many other diseases of swine, some of which are communicable in a certain degree, but none of which are so widespread and fatal as the two that have been named. Some of these, especially the parasitic ones, are receiving attention, and may in the future be treated at length in the bulletins of this Bureau.

The greater part of the detailed study of the disease, the planning of experiments, and the bacteriological investigations have been carried

7

out by Dr. Theobald Smith, while the conducting of the experiments, the care of the experimental animals, and the general management of the experiment station have been under the direction of Dr. F. L. Kilborne. I can only speak in the highest terms of the untiring industry and skill displayed by both of these gentlemen.

The experiments in regard to prevention and medical treatment are being continued, but it is confidently believed that an understanding of the facts detailed in this volume will enable the owners of hogs to prevent or even arrest the great majority of outbreaks. It should be understood, however, that the measures indicated must be enforced promptly, energetically, and thoroughly.

D. E. SALMON,
Chief of the Bureau of Animal Industry.

Hon. NORMAN J. COLMAN,
    Commissioner of Agriculture.

THE INTRODUCTION AND SPREAD OF HOG CHOLERA IN THE UNITED STATES.

In the early days of hog-raising in the United States these animals were comparatively free from disease, and wide-spread epizootic maladies among them appear to have been unknown. A circular letter of inquiry was recently sent from the Bureau of Animal Industry to the correspondents of the Department and to swine-breeders in all parts of the country, asking the date of the first appearance of hog cholera in the respondent's county, and a statement as to the health of swine previous to that time. More than a thousand replies have been received, many of them so carefully written as to be of much interest and value. Nearly all agree in stating that at one time the swine industry was not subject to the periodical losses from epizootics which now cause such discouraging losses. From the first appearance of this class of diseases the outbreaks became more numerous and more wide-spread until nearly all parts of the country are now subject to frequent invasions.

The correspondence on this subject can only be briefly summarized in this bulletin. The first outbreak of disease supposed to be cholera that was referred to occurred in Ohio in 1833. It is reported from one county in South Carolina in 1837, and from one in Georgia as having existed in 1838. It existed in 1840 in one county in Alabama, one of Florida, one of Illinois, and one of Indiana. In 1843 it is reported from one county in North Carolina. In 1844 one county in New York reports being affected. Its presence in 1845 is only mentioned by one correspondent from Kentucky.

The thirteen years, from 1833 to 1845, inclusive, form a period in which but ten outbreaks of swine disease, supposed by the writers to have been hog cholera, were mentioned in these replies. It is evident that during this period hogs were generally healthy throughout the country, and that the losses from disease were not sufficient to attract very much attention. The nature of the disease referred to as existing so long ago may, of course, be questioned at this day, and we have no means of deciding whether or not any particular outbreak was cholera or some non-contagious malady due to local conditions. It is reasonable to conclude, however, that the correspondents are correct in their opinion in most cases, because since 1845 the outbreaks mentioned became more numerous year by year until we find nearly the whole

9

country involved. This rapid increase of the number of affected sections would seem to indicate that a contagious disease had been introduced and carried to widely separated sections of the country, from which it extended until, with a year favorable to its propagation, we find a sudden and alarming increase.

Turning again to the number of outbreaks reported, we find, in 1846, that there were two in North Carolina, one in Georgia, one in Illinois, one in Indiana, and one in Ohio. In 1847 four are given in Tennessee and one in Virginia. In 1848 we hear from it in one county in Illinois, two in Indiana, two in Kentucky, one in North Carolina, and one in Virginia. In 1849 it is mentioned as in one county in Indiana and in one county in Ohio. In 1850 we have accounts of three outbreaks in Georgia, one in Pennsylvania, one in Indiana, two in Kentucky, one in North Carolina one in Ohio, and two in Tennessee. In 1851 there were outbreaks in Illinois, Indiana, and Tennessee. In 1852 there were two reported in Illinois, two in Indiana, one in Missouri, and one in Ohio. In 1853 it invaded two new counties in Illinois, two in Indiana, two in Kentucky, one in North Carolina, four in Ohio, two in Tennessee, one in Texas, and one in Virginia. In 1854 it appeared in one more county in Illinois, four in Indiana, five in Kentucky, two in North Carolina, two in Ohio, and one in Tennessee. In 1855 it is found in six counties in Illinois, five in Indiana, one in Kansas, four in Kentucky, one in Missouri, two in Tennessee, and one in Virginia.

The number of outbreaks mentioned by correspondents, it will be seen, is not less than ninety-three for the ten years from 1846 to 1855 inclusive. As compared with the ten outbreaks reported for the previous thirteen years this is an enormous increase. There can be little doubt that it was during the period from 1846 to 1855 that hog cholera became scattered over the country and fairly began that work of destruction which has become so familiar to us in later years.

Below will be found a tabulated statement of the replies from nearly eight hundred and fifty of our correspondents. The figures show the number of original hog-cholera infections reported for the different periods from the first recorded appearance of the disease in this country to 1887. Of course there have been many counties infected within that time which are not referred to in these communications, but the large number that were mentioned gives as perfect an idea as can now be obtained of the development and spread of this contagion. It is to be remembered that the outbreaks mentioned are not secondary infections, but are the first outbreaks of the disease in the correspondent's local. ity, and in most cases the first which occurred in his county. In nearly all cases it is stated that previous to the outbreak referred to the health of hogs had always been good, and the losses from disease had been confined as a rule to a single animal at a time.

Number and dates of original infections with hog cholera and swine plague, as compiled from recent correspondence.

States.	Thirteen years, 1831–'45.	Ten years, 1846–'55.	Five years, 1856–'60.	Five years, 1861–'66.	Five years, 1866–'70.	Five years, 1871–'75.	Five years, 1876–'80.	Five years, 1881–'85.	Two years, 1886–'87.
Alabama	1		4	7	5		1	1	
Arkansas		1	2	7	1	4	2	2	
California									2
Colorado								1	1
Connecticut								2	1
Dakota						1		1	1
Florida	1			1		1	3	3	
Georgia	1	5	13	13	7	4		1	1
Illinois	1	14	40	14	7	8	7	1	
Indiana	1	19	24	8	10	5	4	3	
Iowa			7	15	11	11	17	13	3
Kansas		1	2			5	12	20	6
Kentucky	1	15	18	7	4			3	2
Louisiana			2	3	3		1	1	
Maine							1		
Maryland			1	2				3	1
Massachusetts				1				2	2
Michigan							2	3	3
Minnesota						1	1		
Mississippi			4	2	3		1	2	2
Missouri		2	6	8	16	8	8	2	1
Nebraska			1		1	2	3	15	3
New Hampshire								1	
New Jersey			2	1				1	1
New York	1				1			2	
North Carolina	1	7	4	10	9	3	4	2	
Nevada								1	
Ohio	1	11	7	6	9	12	7	1	9
Pennsylvania		1	3	2		2	4	3	1
South Carolina	1	1	3	1		1			
Tennessee		12	16	10	5				1
Texas						2	7		2
Virginia		4	2	5	3	8	4	5	3
West Virginia			2	2	4		4	1	1
Wisconsin			1		1		4	13	2

Whether the outbreak which occurred in Ohio in 1833 was the first introduction of hog cholera in this country or not, can not now be determined. It seems reasonably certain, however, that the contagion was imported from Europe with some of the animals that were brought from there to improve our breeds of swine. The investigations made in England and on the continent during the last year demonstrate that the swine fever of Great Britain is identical with our hog cholera, and that this disease is also widely scattered over the continent of Europe. This being the case, it would appear much more likely that the contagion was imported from there, as we know occurred with the contagion of pleuro-pneumonia of cattle, than that it appeared spontaneously or was developed by the conditions of life in this country. Having been once introduced it spread gradually, following the lines of commerce and being for a long time confined to them, until, extending step by step, it has at one time or another invaded every section of the country in which swine raising is a prominent industry.

Dr. George Sutton of Aurora, Ind., in 1858, wrote as follows:

I have seen notices of this disease prevailing in the States of Illinois, Kentucky, Indiana, Ohio, New York, Massachusetts, Pennsylvania, and Maryland. It has prevailed extremely in Indiana, particularly in Dearborn, Ohio, Ripley, Rush, Decatur, Brown, Bartholomew, Shelby, Johnson, Morgan, Marion, Boone, Posey, and Sullivan Counties. It has also prevailed in Campbell, Kenton, Boone, Gallatin, Carroll, Breckinridge, Bullitt, Bath, Henry, Henderson, Nicholas, Livingston, Union, and Crittenden Counties, Kentucky. It has also prevailed in Hamilton, Butler, Clinton, Fayette, and Clermont Counties, Ohio. Also in different portions of Illinois, and very severely in Wayne, White, and Gallatin Counties. It has also prevailed in the State of New York. The Ohio Farmer for January 3, 1857, quoting from the Buffalo Republic the extensive prevalence of the disease, says that "in western New York especially we learn it has been very fatal, but is now over. In conversation with one of the most extensive dealers in the neighborhood, a day or two since, he informs us that about six weeks ago he lost about four hundred in a very short space of time. A distiller in Jordan, during the month of September, lost fourteen hundred, which cost, in addition, over $1,000 to have them buried. In Rochester, at all the principal points, and even among the farmers, the mortality has exceeded anything ever before heard of. A butcher in this city not long since purchased $500 worth of fat hogs, but they died so rapidly on his hands that he scarcely realized $75 on the investment." The Worcester (Mass.) Spy reports that many farmers in that city and vicinity are losing their swine by the mysterious and fatal disease known as the hog cholera. In the southeast part of the town it prevails in a greater or less extent upon nearly every farm.

In most cases the disease is traced to Western hogs that have been sold by the drivers the present season, and which seem to have communicated the contagion to the other inmates of the sties in which they have been kept. It is known that of many droves of Western shoats that have been sold at Brighton this season, and peddled about the State, nearly all have died. The disease has, no doubt, prevailed extensively in other parts of the country, of which I have seen no notice. In this section of the country it has been extremely fatal. Over portions of Dearborn County it spread from farm to farm, and some of our farmers lost from 70 to 80 out of 100 of their hogs. At the distilleries the mortality has been very severe. I received information that more than 11,000 died at the distillery in New Richmond, in the summer and fall of 1856. The owners of the distillery at Aurora inform me that they have lost between 6,000 and 7,000. A gentleman informs me that he lost in 1856, at Ingraham's distillery, in Cincinnati, from the 1st of August up to the 24th of October, 1,285, losing 1,152 out of a lot of 2,408. Another gentleman informs me that at the distillery in Petersburgh, Ky., he lost from the 1st of June up to the 18th of October, 1856, 2,576. I have also received information from several other distilleries where the losses were large.*

According to Dr. Sutton, this disease first appeared in ·Dearborn County, Ind., in July, 1850.

* George Sutton, M. D., Observations on the supposed relations between epizootics and epidemics, and experimental researches to ascertain the nature of the recent epizootic among the swine, and the effects which diseased meat may have on human health. The North American Medico-Chirurgical Review, May, 1858, pp. 483–504.

Dr. E. M. Snow writes that—

During the last five years this disease has been seen, from time to time, in portions of the more eastern States, sometimes, as in western New York in 1856, proving quite severe and fatal in comparatively limited localities. But in the Eastern States it has, to a great extent, originated with and has generally been confined to hogs imported from the West. I think that in no State east of Ohio has the disease prevailed extensively, or attained the character of a wide-spread epidemic.

In the vicinity of Providence, R. I., it has prevailed to some extent, more particularly among large herds of swine during each of the last five winters, but has been mostly confined to hogs brought from the West, and has usually disappeared with the approach of warm weather. During the last winter it was more severe than in any preceding, and was not confined to Western hogs. Neither did the disease, as heretofore, cease with cold weather, but it continued until August, having destroyed more than 500 hogs in Providence and in the adjoining towns during the first seven months of the present year, 1861. I have also heard of its prevalence in various towns in Massachusetts during the same period.*

The losses from hog cholera in the United States have been enormous. Estimates have from time to time been made from carefully compiled data, and these have, so far as the writer is aware, never been less than $10,000,000, and have reached $25,000,000 annually. The inclusion of losses from other diseases is, however, unavoidable in such estimates, and consequently some allowance must be made for these. The recent identification of an epizootic pneumonia of hogs by the Bureau of Animal Industry, a disease which appears to be identical with the *Schweineseuche* of German writers, shows that the varieties of swine diseases in this country are more numerous than has been supposed. The erysipelas of Europe (French, *rouget;* German, *Rothlauf*), and charbon have not yet been identified as occurring in an epizootic or enzootic form among swine in the United States but the existence of these diseases is not impossible, as the investigations have not yet been sufficiently numerous to reveal the nature of all such outbreaks. The diagnosis of such diseases has been very uncertain in the past, because the symptoms were not clearly defined, and not always sufficiently characteristic. The most reliable means of discrimination between these maladies at present is the isolation of the microbes which produce them. The characteristics of these organisms are now so well known that the bacteriologist has no difficulty in distinguishing between them.

* Edwin M. Snow, M. D., Hog Cholera. Annual Report U. S. Department of Agriculture, 1861, p. 147.

THE INVESTIGATIONS OF SWINE DISEASES.

Among the first to investigate these diseases in the United States from a medical point of view was Dr. George Sutton, of Aurora, Dearborn County, Ind. In the extensive epizootics which prevailed in that county from 1850 to 1858 he had abundant material for study, and he noted the more salient features of the plague in a very clear and comprehensive manner. While it is extremely difficult at this day to decide whether the outbreaks studied by Dr. Sutton were hog cholera or swine-plague, or a combination of the two, his observations are very interesting and bring out many important facts. Of the symptoms, *post mortem* appearances, and nature of the malady, he says:

This disease presents a great variety of symptoms. In January, 1856, when this epizootic was at its height in this section of country, I published a short notice of it in the Cincinnati Gazette. The symptoms which I then described I have found upon a more extensive observation to be correct. The hog first appears weak, his head droops, and sometimes in a few hours after these symptoms diarrhea commences. There is frequently vomiting. In some cases the discharges were serous and clay-colored, sometimes dark, also bloody, and mucus resembling those of dysentery. The urine at first was generally small in quantity and high-colored, but as the animal recovered it became abundant and clear; this was one of the symptoms by which the men who were attending the hogs at the distillery ascertained that they were recovering.

In a large number of cases the respiratory organs appeared to be principally affected; there was coughing, wheezing, and difficult respiration. In some instances the animal lost the power of squealing, and the larynx was diseased. There was frequently swelling of the tongue and bleeding from the nose. In those cases where the respiratory organs were the principal seat of the disease there was generally no diarrhea or dysentery. In many instances the disease appeared to be principally confined to the skin; sometimes the nose, the ear, or the side of the head were very much inflamed; the ear swollen to twice its usual thickness. This inflammation would spread along the skin, sometimes over the eye, producing complete blindness. Sometimes one or more legs were inflamed and· swollen, and the inflammation also extended along the body. The skin where w as inflamed was red and swollen. Some had large sores on their flanks or sides from 3 to 6 inches in diameter. In one instance, at the distillery, the inflammation extended along the fore leg, the foot became ulcerated and sloughed off, and the animal recovered. Some appeared delirious, as if there was inflammation of the brain. I examined the blood of four hogs which had this disease well marked; they were stuck, and the blood, arterial and venous, was caught in a bowl. It was cupped and presented a well-marked buffy coat. Death took place in from one to ten days after the attack. Sudden changes in the weather, particularly from warm to cold, appeared to increase the fatality of this disease. The average mortality of hogs that were in pastures or fed on slop was from 33 to 45 per cent., but it was frequently much more fatal if hogs were fed on corn—in some instances ranging from seventy to eighty out of the hundred, and in some instances even higher.

I found on opening the bodies of hogs that had died of this disease that they all presented evidences of a diffusive form of inflammation. From sixty-seven hogs that I have examined I found it was not confined to any particular tissue. Sometimes this inflammation was confined to one organ; in other cases it attacked several at the same time. The skin frequently presented patches of inflammation, and often had a purple appearance. In cutting through parts that were the most inflamed, the skin was swollen and the cellular tissue was infiltrated with serum. Frequently, however, the skin was merely discolored, without any swelling whatever. The stomach was occasionally distended with food, and the mucous membrane in nearly every instance presented evidence of inflammation, sometimes extending over the whole stomach, at others only in patches. It was generally of a deep-red color, thickened, and frequently softened. Sometimes it was covered with a viscid mucus; in other instances there was an effusion of blood into the stomach. The mucous membrane of the small or large intestines, where there had been diarrhœa or dysentery presented in all instances evidences of inflammation; in patches it was red, thickened, sometimes softened, and occasionally ulcerated; where there had been dysentery there was generally bloody mucus found in the large intestines. The bladder generally contained urine; sometimes its mucous membrane was inflamed, and in one instance there was an effusion of blood into this organ. In a large number of cases I found evidences of peritoneal inflammation, such as redness of this membrane, effusion of turbid or bloody serum, adhesions between the intestines and between the intestines and sides of the body. In three instances blood was effused into the peritoneal cavity—in one instance more than a quart; it appeared in this case to come from the liver. The liver was occasionally the seat of this inflammation, not only in its investing membrane, but the parenchyma; sometimes there were abscesses, and in one instance portions of it were gangrenous. The lymphatic glands were generally of a dark-red color, frequently resembling clots of blood. This disease of the lymphatic glands was of common occurrence.

The lungs were frequently the seat of this inflammation, portions of one or both presenting different appearances, from simple congestion to complete hepatization; sometimes there was ulceration, and frequently there was a turbid or sero-purulent or bloody effusion into the pleural cavity; sometimes there were extensive adhesions between the lungs and pleura of the ribs. At first I was inclined to believe this malady to be a form of pleuro-pneumonia, but after I became better acquainted with it I found that the inflammation was not uniformly confined to any organ. In a number of instances the mucous membrane of the bronchia was deeply inflamed and the inflammation extended to the trachea and larynx. In several instances the larynx was inflamed, resembling laryngitis. One animal that had great difficulty in breathing and could make no noise I had knocked in the head, and on examination I found the mucous membrane of the larynx and epiglottis inflamed and swollen; also the tongue was swollen. There were evidences in several instances of pericarditis, which had produced adhesions between the heart and the pericardium. The brain, from the difficulty of opening the skull, was examined only in one instance; it was found healthy, although I feel confident it was frequently the seat of the disease.

From these examinations we see that it is a misnomer to call this malady cholera. It is a contagious inflammatory disease, the inflammation being confined to no particular tissue, sometimes attacking only one, at others several, in the same animal. Evidences of this inflammation were found in the dermoid, the cellular, the serous, the mucous, and glandular tissues. I consider it a diffusive form of inflammation from the manner in which I have witnessed it spread along the skin. In one night I have seen it extend from the eye to the ear—the ear becoming inflamed and swollen.

Although we have not been able to show that this is a cholera epizootic, still the facts elicited may be of interest and remove doubts at some future period. But, then, if this malady does not resemble cholera, does it resemble any of the diseases to which the human system is subject? I think not. Like the specific eruptive diseases, it is highly contagious; the infection has a period of incubation of from twelve to twenty days, and one attack appears to exempt the animal from a second. But in this disease, although petecchiæ and an eruption may appear on the skin, its principal characteristic is a diffusive form of inflammation which may attack nearly every tissue and spread like an erysipelas. But then, again it differs from this disease, as it is well known that in erysipelas one attack does not exempt the system from a second; and although erysipelas may be contagious, still it is doubtful whether the period before the eruption shows itself is so uniform as in this disease, and while erysipelas is generally confined to the skin this inflammation most frequently attacks the lungs and mucous membrane of the alimentary canal. This disease appears to be intermediate between the specific eruptive diseases and erysipelas, partaking of the nature of each, and probably not having its exact resemblance among the diseases to which the human system is subject.*

Of his experiments and observations on the contagiousness of the disease and the manner of its spread he writes:

When the disease made its appearance in this section of country, in the summer of 1850, and we saw it spreading from farm to farm, the question was suggested whether it did not spread by contagion, for it was not known at that time (July) that the malady was contagious. Feeling much interest in watching the progress of this disease, from the large numbers of animals that were dying, I suggested to the owners of the distillery at Aurora, Messrs. Graffs, that we should try a series of experiments to ascertain the nature of the disease, and whether it was propagated by contagion. To this they readily assented, and as they were constantly receiving fresh hogs, there was a fine opportunity to make any experiments we saw proper. I am indebted to Mr. J. J. Barkman, of Aurora, for seeing that the following experiments, with the exception of the last, were carefully made. The hogs on which the experiments were made were known to be healthy:

(1) Six hogs that had been exposed to the malady by being in contact with diseased hogs were put into a yard by themselves and fed on slop and corn; on the fourteenth day from the time they were exposed to the disease they were all unwell; 3 died within a week afterwards, the rest recovered.

(2) Ninety hogs were exposed to the disease, then put into a yard by themselves and fed on corn and water (no slop given); in thirteen days disease made its appearance among them, and they continued to die until 60 out of the 90 died.

(3) Fifty hogs were put into a pen by themselves, and fed on slop; they had not been exposed to the disease; for six weeks they continued healthy.

(4) One hundred hogs that had not been exposed to the disease were put into a yard by themselves, and fed on corn and water; for thirty days no symptoms of disease appeared among them. They were then put into a pen with diseased hogs; on the thirteenth day they began to show symptoms of the malady, and the disease rapidly spread among them, until 40 died.

(5) Thirty-three hogs, out of a lot of 263, were put into a pen by themselves; for six weeks they continued healthy. The remaining 230 were put into a pen adjoining in which were diseased hogs; in thirteen days disease made its appearance among them, and continued until one-half died.

(6) Four young and healthy hogs were put into a pen in which, four days previous, diseased hogs had been; they were fed on corn and water. On the fourteenth day they were all unwell; and one died on the fifteenth day, and in five days more they were all dead. This experiment shows that the infection may be retained in a pen several days.

* North American Medico-Chirurgical Review. 1858, II, 496.

(7) I inoculated, on the 2²th of October, five healthy hogs with the blood taken from the inflamed tissues of hogs that had died of this disease. On the fourteenth day (November 11) they were all unwell, and all died with the exception of one. In three inflammation spread from the point where they were inoculated, along the skin and down the legs, which became very much swollen. I can not say that this inflammation was caused by the inoculation, for it did not appear until the fourteenth day, and many hogs had this external form of disease.

From these experiments I think that we not only ascertained that this disease was infectious, but that the infection had a latent period of from twelve to fifteen days. Observations have since led me to consider the latent period as varying from twelve to twenty days. These experiments also showed that the hogs in the pens were not dying from strychnine in the slop. A statement was going the rounds of the papers about this time that strychnine was used in making yeast at the distilleries, and was poisoning the hogs at these places in large numbers.

The manner in which this disease in many instances spread among hogs from farm to farm also showed most conclusively that it was infectious. One farmer had 75 hogs that he turned into a corn-field to fatten. These hogs had been exposed to the disease, had become sickly, and numbers had died. He bought 36 more; these were all healthy; they were put into the same field with the diseased hogs: in two weeks they were unwell, and numbers died. He bought 45 more, all healthy, and put them in the same field; in two weeks they began to show symptoms of disease, and in a few days after a number died. Finding that he was likely to lose all his hogs, he sold 50 of the fattest that were left to the butchers in Cincinnati, at a reduced price, as diseased hogs. These hogs, it was said, were purchased for their fat, to be used in the manufacture of lard oil. After deducting the 50 he sold, he lost out of the 156 all but 10.

A large number of facts could be given, if necessary, to show the contagiousness of the disease. Hogs that had been once under the influence of this malady appear not to be susceptible to a second attack. Although placed in pens with diseased hogs they continue healthy and fatten, and not a single instance can I find of hogs having this disease twice. There is reason to believe that this malady occasionally prevails at the distilleries in a mild form, as the old hogs at many of these places did not take the disease. But why it should suddenly have assumed so malignant a character is as difficult to account for as that scarlatina, from being at one time the mildest of diseases, at another becomes one of the most malignant. On the farms the old and young hogs appear to be equally susceptible to the disease.

In 1861 Edwin M. Snow, M. D., of Providence, R. I., contributed a paper on hog cholera to the Annual Report of the United States Department of Agriculture (pp. 147–154). The symptoms of the disease he gives as follows:

The symptoms, as described by persons unaccustomed to such observations, are extremely various. By combining the information obtained from others with results of our own observations the symptoms, as seen during the life of the animal, are nearly as follows:

(1) *Refusal of food.*—This is the first symptom usually noticed by those who have the care of the animals, though, as will be seen hereafter, this symptom by no means indicates the beginning of the disease. The refusal of food, after it is first noticed, generally continues through the whole sickness, and food of every description is mostly refused.

(2) *Great thirst.*—This is constant, and large quantities of cold water will be swallowed if it can be obtained. Even after the animal is unable to stand alone it will drink cold water with eagerness.

(3) After a time, the length of which varies very much, the animal begins to show signs of weakness; reels, staggers, and, in attempting to walk, often falls down,

(4) In most cases there is a diarrhea, with copious fluid discharges of dark, bilious, and very offensive matters. In a few cases there is no diarrhea, but evacuations of hard black balls; but in some of these cases the fluid offensive matter is found in the intestines after death.

(5) In a few cases there is vomiting; but this is not often severe, nor is it continuous for any length of time.

(6) The external appearance of the animal is at first paler than usual, but towards the last of the sickness purple spots appear, first on the nose and sides of the head. These extend along the sides of the belly and between the hind legs, after which the animal soon dies.

(7) In many cases, perhaps a majority, ulcers are found on the different parts of the body. These were particularly noticed on the inside of the lips and gums, and on the feet, and were often quite deep and excavated. In some cases these ulcers were seen in the nostrils, and in one case there were extensive ulcerations in the back part of the mouth, on the tonsils.

(8) In some cases the legs are swelled, and the animal is lame; sometimes the ears and sides of the head are swelled and red; sometimes the eyes are sore and inflamed; sometimes swellings like carbuncles are seen; and, generally, the glands near the surface seem to be enlarged.

(9) In most cases the pulse is quickened, the breathing is hurried and difficult, and there is much cough. But in some genuine cases there is no perceptible trouble with the lungs, and no important signs of disease are found in them after death.

(10) The duration of the disease in fatal cases, after the first symptoms are noticed, is extremely variable. We have seen some which have died within two or three hours; others have lived many days. It is difficult, however, to fix the time of the appearance of the first symptoms. The first noticed is usually the refusal of food; but it is probable, indeed it is certain, that the sickness is in progress for a considerable period before the animal refuses food. Cases like the following are sometimes seen: A hog refuses to eat; it soon grows weak; staggers in walking; turns purple on the sides and belly, and dies within two or three hours after the first symptom is noticed. But, on examination after death, extensive disease is found in the intestines, or in the lungs, or in both, at a stage of development which must have required many days to reach.

The *post-mortem* appearances are then summarized in the following language:

Having described the symptoms, as seen while the animal is living, I will now give, briefly, the appearances found on examination of the bodies after death:

In the course of our investigations, during the last winter and spring, the bodies of nine hogs were carefully examined by Drs. G. L. Collins, J. W. C. Ely, and E. T. Caswell, of Providence, in the presence of several other physicians. A minute account of each case was prepared by Dr. Collins, and published in the Transactions of the Rhode Island Medical Society for 1861. It will be sufficient for the objects of this paper to give a brief synopsis of the diseased appearances which were found in these examinations:

Lungs.—In two cases the lungs were healthy. In the remaining seven cases one or both lungs were more or less inflamed, having a liver-like appearance, called hepatization. In some cases the inflammation was more advanced, and the substance of the lungs was breaking down into a mass of disease. In all cases where the lungs were inflamed there was also pleurisy, and the lungs were adherent to the walls of the chest; the inflammation of the lungs and the pleurisy together constituting true pleuro-pneumonia. In two cases there were tubercles, or consumption in the lungs; but this is not uncommon in hogs, and is not supposed to have any connection with the special disease we are considering.

Stomach.—The stomach and the small intestines were generally healthy. The stomach was frequently distended with an offensive mixture of food, and in one case the inner surface was ulcerated to some extent. In two cases worms were found in small intestines, but this was probably a merely accidental occurrence, and had no necessary connection with the disease.

Large intestines.—The inner coat of the large intestines was generally inflamed and softened with ulcerations to a greater or less extent, and they were frequently so tender as to be easily torn with the fingers. On account of their diseased condition their inner coat was frequently discolored. The inflammation and ulceration of these intestines are probably the principal cause of the diarrhea in this disease.

Kidneys.—These organs were, in every case, much more pale and yellow than natural: this condition was well marked.

The *liver* and *bladder* were generally healthy. In some cases water was found in the cavity of the belly and of the chest, and in the membrane surrounding the heart (heart-case). In two cases numerous minute purple spots were seen upon the membrane lining the walls of the belly. The urine was often changed from the healthy condition, containing albumen and other diseased products, not, however, noticeable to the eye. Ulcers upon the feet and in the mouth were often found. The brain was not examined, as there were no symptoms observed which seemed to indicate disease of this organ. It may be at times affected, but is probably more rarely so than the other organs of the body.

Such are some of the most important appearances which are found on examination of the bodies of hogs which have died with this disease. It will be noticed that three of the diseased conditions I have described are prominent, important, and such as would be readily recognized by the most ignorant observer.

These are, first, the pleurisy and inflammation of the lungs; second, the inflammation, ulceration, and softening of the inner coat of the large intestines: and, third, the pale and yellowish color of the kidneys. One or more of these diseased conditions will be found in every case, and in perhaps a majority of cases they will all be found in the same animal.

While Dr. Snow admitted the disease to be epizootic he did not consider it to be contagious. Indeed, he neither considered epizootic nor epidemic diseases to be contagious, but, on the contrary, held that they were caused by (1) "an epidemic atmospherical poison," and (2) "the local conditions or circumstances adapted to receive and propagate the poison existing in the atmosphere."

In 1875 Prof. James Law contributed a paper to the report of the Department of Agriculture giving his observations upon hog cholera, or "intestinal fever in swine," in which he so completely embraces the knowledge of our swine epizootics as it then existed that his paper is here reproduced in full. It is as follows:

Synonyms.—Typhoid fever, enteric fever, typhus carbuncular fever, carbuncular gastro-enteritis, carbuncular typhus, pig distemper, blue sickness, blue disease, purples, red soldier, anthrax fever, scarlatina, measles, diphtheria, erysipelas.

Definition.—A specific, contagious fever of swine, characterized by congestion, exudation, ecchymosis, and ulceration of the mucous membrane of the intestines, and to a less extent of the stomach; by general heat and redness of the skin, effaceable by pressure; by small red spots, complicated or not by elevations and blisters; by black spots and patches of extravasated blood on the integument, the snout, nose, eyes, mouth, and all other visible membranes, and on internal organs, ineffaceable by pressure and tending to sloughing; usually by liquid and fetid diarrhea; and by a very high and early mortality.

The malady has been long known to pig-raisers and pork-factors in the Old World and the New, but in veterinary works it has been mistakenly placed in the list of malignant anthrax affections, to which many of its lesions bear a striking resemblance. Two English works, published within the last year, repeat this time-honored fallacy. To malignant anthrax it is allied by the altered condition of the blood, by the solution of the blood-globules, by the imperfect coagulation of the blood in many cases, by the occasional enlargement and congestion of the spleen, by the extravasation of the blood out of the vessels (petechiæ) into the skin, mucous membranes, and internal organs, and by the dusky hue of the eye, nose, etc. That it is essentially distinct is shown by the fact that its virus, so frightfully contagious and fatal to pigs, is not communicable to any other domestic animal. The contagion of malignant anthrax, on the other hand, is deadly to all domestic animals, and even to man himself.

The common American designation of hog cholera has only the diarrhea to support it, and, as we see outbreaks in which this feature is mainly remarkable for its absence, the name comes to be an absolute misnomer. In many cases a gelatiniform exudation takes place on the affected surface of the mucous membrane of the intestines, windpipe, or bronchia, and the disease has accordingly been named diphtheria. But as such an exudation is by no means constant, a name founded on this peculiarity would have no actual foundation in a large proportion of cases. Again, the exudation (see *Post mortem* No. 1) is mainly composed of cells and granules, with less of the fibrinous matrix than is usual in diphtheria. Lastly, the intestinal fever of swine is most virulently contagious, whereas diphtheria is very slightly, if at all, infectious, and is confined rather to certain insalubrious localities or buildings.

From scarlatina and measles it is sufficiently distinguished by the constancy of the intestinal legions, though it resembles both in its cutaneous rash. With erysipelas it has no real connection, the one common feature, the redness of the skin, being due to a condition altogether different in nature, progress, and results.

The constancy of the congestion, specific deposit, and ulceration in the intestines in the fever of swine, characterize it as perfectly as do the same lesions in typhoid fever in man. It further agrees with typhoid fever in having a higher evening than morning temperature and a rose-colored eruption on the skin. To this disease, indeed, it bears a closer resemblance than to any other disorder of man or beast, so that Dr. Budd and others with much plausibility call it the typhoid fever of pigs. But in spite of the similarity of the specific deposits and ulcerations on the intestines, those of the pig show less tendency to attack the agminated glands (Peyer's patches) and the solitary glands than is the case in man. They appear on all parts of the mucous membrane of the large and small intestines, yet the agminated and solitary glands rarely escape entirely, and sometimes they alone are the seats of ulceration and morbid deposit. The skin eruption, too, in the pig-fever is often distinctly raised, and even vesicular, whereas that of typhoid fever is a simple rash, and, like a blush, may be completely though temporarily effaced by pressure. Finally, the contagion is incomparably more virulent and tenacious of life than that of typhoid fever, and the mortality is greater and occurs earlier in the disease. On the whole, we must look on this affection of pigs as a disease *sui generis*, having close affinities with the typhoid fever of man, yet essentially distinct, and hence the term intestinal fever of swine is more applicable, as at once expressing its nature and avoiding confounding it with other and distinct affections.

Incubation.—The period of incubation has not been definitely settled. My experience in Scotland in 1864 would have led me to set it down at from seven to fourteen days. The infected pigs were four days on the journey from the English market by rail and seven days on the farm before the disease manifested itself. Again, the home-bred swine were sound until four weeks after the strange hogs came on the farm, and three weeks after the latter were generally sick. Pigs, though farrowed by sick dams, did not show any sign of disease for about a week, although nearly all eventually died.

In Dr. Budd's cases, in April, 1865, the first symptoms of illness were shown four or five days after the pigs were brought from Bristol market, where they may or may not have been infected.

Dr. Sutton's experiments, made at Aurora, Ind., in September, October, and November, 1842, deserve repetition in this connection. (1) Six hogs, after contact with diseased animals, were placed in a sound pen, and sickened on the fourteenth day. (2) Of 90, similarly exposed and then put in a sound yard, some sickened on the thirteenth day. (3) One hundred, similarly exposed, contracted the disease on the thirteenth day. (4) One hundred and thirty, placed in a yard adjoining one occupied by diseased hogs, became ill on the thirteenth day. (5) Four young and healthy pigs, placed in a pen occupied four days previously by diseased hogs, sickened on the fourteenth day. (6) Five healthy hogs, inoculated with the blood from the inflamed tissues of diseased swine, were unwell on the fourteenth day.

Further experiments were made by Professor Axe, of London, in April, May, June, and July, 1875. (1) Two healthy pigs were (in April) placed for forty-eight hours in the same house with a diseased one, contact being carefully avoided. One was dull and off its food on the sixth day and the other on the eighth. (2) On May 15 a pig was inoculated with the liquid cutaneous exudation, which had been kept on dry ivory points for twenty-six days. On the fifth day there was slight dullness and heat of skin, and on the sixth the malady was well developed. (3) On June 10 another pig was inoculated with the cutaneous exudation of the last, the operation being performed by another party and the pig kept apart to avoid all risk of indirect contagion. On the fifth day temporary redness was noticed on four teats, and on the sixth day the symptoms were fully developed. (4) Another pig broke into the pen occupied by the last-named subject and was left there for six days, when it was taken out seriously ill. In the hot summers of Illinois instances are met with in which symptoms of the disease are shown in a previously healthy herd under three days after the wind has blown from the direction of a sick lot half a mile distant. In analyzing this apparently somewhat discordant evidence we must bear in mind that the period during which a poison will remain latent in the system will vary according to the amount taken in, the excited or febrile condition of the subject, and the mode of introduction into the system. Thus an excess of any poison, animal or vegetable, will usually show its effects with remarkable rapidity. A feverish state of the system, whether induced by intense heat, passion, or disease will rouse the poison into unusually early activity. Lastly, poisons that are inoculated usually act sooner than those introduced into the system by other channels. These considerations will serve to reconcile the prolonged latency of the poison in Dr. Sutton's cases, observed in cold weather, as compared with Dr. Budd's, Professor Axe's, and my own, in the English summer, and of these in their turn with the prompt development of the malady in the semi-tropical summer of Illinois.

Symptoms.—The earliest symptoms are slight dullness, with sometimes wrinkling of the skin of the face as if from headache; shivering or chilliness and a desire to hide under the litter are not uncommon. Some loathing of food, intense thirst, elevation of the temperature of the rectum to 104° Fahrenheit and increased heat and redness of the skin are usually the first observed symptoms, and occur one or two days later than premonitory signs above mentioned. The increased heat of the skin is especially noticeable inside the elbow and thigh and along the belly. By the second day of illness the whole surface feels hot, and in white pigs is suffused with a red blush, which may pass successively through the shades of purple and violet. It is usually more or less mottled at particular points, and a specific eruption appears as rose-colored spots of from 1 to 3 lines in diameter, sometimes slightly raised so as to be perceptible to the touch, and either pointed or more frequently rounded. The redness fades under the pressure of the finger, but only to re-appear immediately. The eruption is usually abundant on the breast, belly, and haunch, the inner side of the forearm and thighs, and the back of the ears. It stays out for two or three days, and may be followed by one, two, or more successive crops of the same kind. The cuticle is sometimes raised in minute blisters, a feature which distinguishes this from the rash of typhoid fever, and the liquid of such blisters inoculated on other pigs communicates the disease. In addition to the rash and simultaneously with it, or soon after, there appear on the skin numerous spots of a dark-red or black color, varying in size from a line to an inch in diameter, on the color of which pressure has no effect. These are due to the extravasation of blood or of its coloring matter from the blood vessels into the tissues, and they dry up into thin scabs or sloughs if the animal survives. Similar petecchial spots appear on the muzzle, in the nose, and on the eyes, and in some instances they are confined to these parts. The tongue is covered by a brownish fur.

From the first the animal is sore to the touch, but as the disease develops the handling of the animal is especially painful and causes grunting and screaming. The pig lies on its belly, and, if compelled to rise and walk, moves stiffly, feebly, unsteadily, and with plaintive grunting. This weakness and prostration rapidly increases, and often ends in utter inability to rise or to support the body on the hind limbs. A watery discharge from the nose, followed by a white muco-purulent flow, is not uncommon. A hard, barking cough is frequently present from the first and continues to the last. Sickness and vomiting may be present, but are far from constant. The bowels are often confined at first, and in certain cases, and even in nearly all the victims of particular outbreaks, may remain so throughout, nothing whatever being passed, or only a few small black pellets covered by a film of mucus. These cases are quickly fatal. More frequently, however, they become loose by the second or third day, and diarrhea increases at an alarming rate. The passages are first bilious and of a light or brownish yellow when not colored by ashes, charcoal, or the nature of the foo l. But soon they assume the darker shades of green and red, or become quite black and intolerably offensive. In such cases the elements of blood, inspissated lymph, and membranous pellicles, sloughed off from the ulcerated surfaces, are usually to be found in them.

The diarrhea becomes more profuse, watery, and fetid; the pulse sinks so as to become almost imperceptible; the cough becomes more frequent, painful, and exhausting; the breathing is more hurried and labored; and the weakness increases until the patient can no longer rise on his hind limbs. At this period the petecchiæ become far more abundant. Before death the animal is often sunk in complete stupor, with, it may be, muscular jerking or trembling, or sudden starts into the sitting posture, and loud screams. In the last stages involuntary motions of the bowels are common.

Exceptionally swellings appear on the flank, with extreme lameness, and extensive sloughs of the skin of the ears or other parts. Palpitations of the heart also occasionally occur as precursors, attendants, or sequels of disease. If the disease should take a favorable turn, slight causes may make an early and perfect recovery, a complete convalescence being established in three or four weeks. A considerable proportion of the survivors, however, linger on in an unthrifty condition for months, evidently suffering from the persistent ulceration of the intestines or infiltration of the lungs. The mortality often reaches 80 or 90 per cent. of all swine exposed, and in case of a certain number of the survivors recovery brings no profit to the owner.

Post-mortem appearances.—The blue color of the skin becomes deeper and more universal a few hours after death. The fat beneath the skin is colored more or less deeply in points corresponding to the discoloration of the integument. The snout is usually of a deep blue, with ineffaceable black spots (petecchiæ). The membrane lining the eyelid, and to a less extent the skin, present similar black spots of extravasation.

The most constant changes are in the mucous membrane lining the alimentary canal, and especially that of the terminal portion of the small intestine (ileum) and the commencement of the large (cæcum, colon). The tongue is furred, but deep red, even eroded, at its base, and the pharynx and adjacent parts usually studded with petecchiæ. The cavity of the abdomen generally contains a few ounces of reddish serum, which coagulates on being heated. The stomach may show no more than a pink blush, but more commonly it is of a deep red, from congestion, especially toward the pylorus, and its mucous membrane is often black throughout from the close aggregation of petecchiæ. The small intestines are usually extensively congested, and of a deep red, in many cases approaching to black, as examined externally. Their mucous membrane in such parts is equally high colored, studded with petecchiæ, and in some cases lined by a firm, semi-fibrinous exudation. A more constant condition is the presence of minute sloughs or erosions in the seat of petecchiæ, and of equally small elevations, due to excessive cell-growth, beneath the epithelium. These commonly have a whitish center, with a yellowish or red border. Such is the appearance in cases that prove fatal within two or three days. In those that have survived longer, extensive ulcers appear, of an inch and upward in diameter, evidently the sequel of the petecchiæ, and especially of the eruption.

These ulcers are often covered by black scabs or sloughs, have irregular projecting margins and a variously colored center, consisting of cells in process of disintegration. They are sometimes situated on Peyer's patches, but show no very marked preference for those over other portions of the mucous membrane. The large intestines present a similar varying vascularity, discoloration, petecchiæ, deposit, softening, and ulceration. The changes are especially marked in the cæcum and colon and in the rectum close to the anus. The solitary glands are often large and open, but the ulcers show no particular preference for the points occupied by them. Extensive extravasations of blood into the coats of the bowels and among their contents are not infrequent, and in certain exceptional and advanced cases the peritoneum is inflamed and false membranes bind the bowels together or to other organs, or to the walls of the abdomen.

In the wind-pipe and air-passages within the lungs, the mucous membrane is usually mottled with black petecchiæ, or covered by a viscid mucous exudation. The anterior lobes of the lungs are often solidified by exudation, but remain bright, red, tough, and elastic (splenisation). Limited hepatization is also exceptionally met with, and even false membranes on the pleura or pericardium.

Petecchiæ are common over the various internal organs—on the lungs, pleura, heart, pericardium, diaphragm, peritoneum, liver, pancreas, kidneys, and bladder. The spleen is large and dark, as is usual in connection with blood poisons. Both sides of the heart contain clots of blood, extending from the auricles and ventricles into the great vessels. In the worst cases, the clot is black, soft, and somewhat diffluent; exceptionally it is firm, and shows a distinct buffy coat. The blood-globules, as seen under the microscope, are more or less puckered or crenated at their edges, and mixed with an excess of granular débris and even in some instances spores of a fungus (*micrococcus*).

As illustrating the various lesions in different cases, I append from my notes of post-mortem examinations two that occurred with two years' interval, near Edinburgh and London, respectively:

I.—No. 4. A three months' old, white Yorkshire pig in excellent condition. Examined a few hours after death, being still quite warm, *rigor mortis* had set in strongly. Along the whole lower surface of the body, from the mouth to the tail, are spots of dark red or purple. On the right side of the head and left side of the chest the spots run into each other, so much that they seem to form a single continuous blush. On the back the spots are smaller and less numerous. There are no spots nor petecchiæ on the snout, but a glairy bloody fluid runs from the right nostril.

The membrane lining the eyelid is congested, having a dark hue, approaching purple, and a portion of the mucous membrane of the rectum, exposed by the relaxation of the sphincter, presents the same appearance. Dark red spots and petecchiæ exist about the vulva. Under the belly is a subcutaneous layer of fat about an inch thick, and beneath the purple spots on the skin this is discolored by blood throughout the entire thickness.

The abdominal cavity contains 6 ounces of a dark bloody fluid. The intestines have a deep florid hue externally. The stomach and the greater part of the rectum are pale and without any lesions of the mucous membrane. Close to the anus this membrane is congested. The stomach contains about two pounds of food (boiled potatoes, corn, etc.). The small intestines have their mucous membrane thickened, soft, friable, and very red, the shades being lighter or deeper at different points. It presents at intervals spots of a much darker hue, approaching purple, and respectively from one to two lines in diameter. The large Peyer's patch on the ileum seems hypertrophied and of a deep red, especially close to the ileo-cæcal valve, where it also shows small ulcers. The large intestines have their mucous membrane of a very bright red, soft, friable, and presenting at intervals small ulcers of about a line in diameter, and corresponding apparently to the solitary glands. Some of these ulcers are of a deep red, and appear on the peritoneal coat as dark spots; others are of a dirty white in the center, with raised red edges, and are not so marked on the peritoneal surface. A small nodule felt through the outer coats is characteristic of both. The contents of the large intestines consist of dark semi-liquid feces, with a great amount of ashes.

The liver is variable in color and very friable. Though still warm, it presents small bubbles of gas at intervals under its capsule and throughout its substances. The gall-bladder is half full of a very light-colored yellow bile. The pancreas seem healthy. The kidneys, bladder, and uterus are normal.

The pleura and lungs appear healthy, excepting a portion of the anterior lobe of the left lung, which is in a state of splenisation. The bronchia contains bloody froth, especially in the left lung, and those of the solidified portion contain a white solid substance, completely filling them, and appearing to the naked eye like a fibrinous clot, while under the microscope it is found to be mainly composed of small globules about the size of blood-cells. It is disintegrated and partly dissolved in a strong solution of potassa. The mucous membrane of the larynx and trachea is congested and the tube filled with a white frothy mucus of an exceedingly tenacious consistency.

II.—No. 2. Three months' old female pig, large of its age. Dead twelve hours. *Rigor mortis* well marked. Skin almost universally of a livid hue, but purple along the abdomen. Back, white. Profuse eruption over the body, but especially abundant on the abdomen. The smallest, and evidently the most recent specimens of the eruption, are individually about a line in diameter, deep purple, and covered by a delicate, slaty-looking skin. The larger spots have a dark, hard, dead center, which appears to spread gradually to the whole of the patch; some appear as a large black scab of one-half to 1 inch in diameter.

Abdomen contains 6 ounces of serum, which forms a solid coagulum on being heated. False membranes bind the large intestines to the lower wall of the belly, also the two horns of the uterus together and to the bladder, and both to the walls of the pelvis. The large intestines are the seat of an exudation half an inch thick, from which a straw-colored fluid escapes on section. The stomach is considerably discolored on its great curvature externally, as if from extravasated blood. The mucous membrane of the stomach presents numerous petecchiæ and ramified redness along the great curvature. The small intestines show slight branching redness on part of the ileum.

Large intestines.—Cæcum has its mucous membrane abnormally vascular; with abundant petecchiæ, and ulcers of considerable standing; these appear as white, raised masses, and have no manifest connection with the solitary glands. The blindgut contains numerous ascarides. The mucous membrane of the colon repeats that of the cæcum, but at one point beneath its serous coat is a blood-clot measuring 1 inch by half an inch, and a quarter of an inch thick.

The liver is healthy. The diaphragm has abundance of petecchiæ on its posterior surface, especially on the cordiform portion, and apparently leading in radiating lines from the center. Beyond the presence of petecchiæ the organs of the chest seem to be little affected.

Causes.—Contagion is the main cause of this disease. The introduction of diseased pigs into healthy herds; the placing of healthy pigs in pens, cars, steam-boats, markets, etc., where diseased swine are then or have formerly been exposed; a fresh breeze from the direction of a diseased herd, though half a mile distant; the passage of men or quadrupeds or birds from the diseased to the healthy; the use of food, litter, or water that has been in near proximity to the affected animals, have each served to transmit the fever.

The virus appears to be concentrated in the bowel-discharges and liquids of the eruption, but doubtless exists in all the liquids and tissues of the body, and is given off into the air from the skin and exposed mucous membranes. Pigs are often born sick, and die in twenty-four hours. The feeding-troughs and water, contaminated by the filthy feet and snouts, are particularly liable to convey the disease.

The malady prevails at all the periods of the year, but it has opportunities for the widest diffusion in dry seasons and countries, where the virulent matter may be dried up and preserved unchanged for an indefinite period, and in this state may be carried by winds and otherwise. Wet weather contributes to the decomposition and destruction of this, as of any other animal poison, but can not influence its propagation by the direct contact of healthy with diseased animals, nor affect its preservation inside dry buildings.

Unwholesome conditions of life contribute largely to its diffusion, if not to its development *de novo*. The malady frequently appears in pigs that have been carried several days in succession in crowded boats or cars, among the accumulated filth of their own bodies and those of their predecessors, and subject to compulsory abstinence from food and water. Again, it will occur in fat hogs that have been driven a number of miles under a hot sun and then cooped up in a filthy, close, ill-ventilated pen, subjected to the reeking fumes of their own bodies and of long-accumulated nastiness. Many think that dry-corn feeding and overcrowding on filthy manure heaps are largely productive of the disease. But it is too much to assume that the poison is developed *de novo* in such conditions. Similar unwholesome influences favor the development of all contagious diseases by loading the blood with effete and deleterious organic matter, and bringing about a feverish and susceptible state of the system. But, on the other hand, such abuse and maltreatment fail, in very many cases, to induce the affection, so that we are left in doubt in regard to those instances in which it appears as to whether the virus was not hidden away in the buildings, cars, etc., and roused into activity by the presence of a fertile field for its growth in the bodies of the pigs, the blood of which was loaded with organic elements in process of decomposition. The important point is this: We know this as a contagious affection, to the propagation of which all probable insalubrious conditions contribute. So soon as we concentrate our attention on this point we have the key to its prevention, if not to its entire extinction. But, while admitting the influence of overcrowding, filth, starvation, and suffering in predisposing to this disease, it ought to be added that the very highest mortality is often reached among pigs kept in the best hygienic conditions, but among which the virus has been accidentally introduced. Again, some hogs, and even families, appear to be insusceptible, and resist the poison which is carrying off all around them. But similar instances of immunity are met with in all contagious affections.

Treatment.—In a fatal contagious disease like that under consideration it is rarely good policy to subject to treatment. The enormous increase of the poison in the bodies of the sick, and the extreme danger of its diffusion through the air, as well as on the feet of men and animals, render the preservation of the victims eminently unsafe and unprofitable. Yet, in the case of very valuable animals, and where seclusion, disinfection, and careful nursing can be secured, it may be resorted to.

A dry, airy, well-littered building may be provided, abundantly sprinkled with a solution of carbolic acid or chloride of lime. Rags steeped in a solution of one or other of these agents may be hung up at intervals, and sulphurous acid set free by burning a pinch or two of sulphur three or four times a day. The dung should be saturated or thickly sprinkled with finely powdered copperas. Any drains will require disinfection in the same way. If the sick animals are kept in the open air, the ground must be freely sprinkled with disinfectants, above all where the dung has been deposited.

The diet should be well boiled gruel of barley, rye, or Graham flour; or, if fever runs high and the temperature is raised by such food, corn-starch made with boiling water or milk may be substituted. Fresh, cool water should be freely supplied, either pure or slightly acidulated with sulphuric acid.

During the early stages, while constipation exists, the bowels may be gently opened by castor oil (2 ounces for a good six-months' pig) or rhubarb (1 dram), aided by injections of warm water. The heat of the skin must be counteracted by sponging with cool or tepid water, as may seem most agreeable to the patient. As the laxative operates, 20 grains of nitrate of potash and 10 grains of bisulphite of soda may be given twice a day in the drinking water. Charcoal may also be given to absorb and neutralize the deleterious products in the bowels. Or the niter may be replaced by any other neutral salt and the bisulphite by another antiseptic agent. If the patient survives the first few days and gives indications of ulceration, by tender abdomen, diarrhœa, and the passage by the bowels of membranous sloughs, oil of turpentine, in doses of 15 or 20 drops, may be given, shaken up in milk or beaten in an egg; or this may be replaced by similar doses of creosote or carbolic acid, or 3 to 5 grain doses of nitrate of silver. It may be necessary to give opium to check excessive purging, or stimulants to sustain the failing strength and very prostrate condition. Infiltrations and inflammations of the lungs and bowels may demand applications of mustard and turpentine to the chest or abdomen. In short, any complication must be combated as it appears, and the soundest judgment will be wanted throughout to adapt the treatment to the various indications. Each case will demand as close attention and as careful an adaptation of remedial measures to its different stages and phases as would a case of typhoid fever in man. In case of recovery a course of tonics (gentian one-half dram, sulphate of iron 10 grains daily) will often be beneficial, and the return to ordinary feeding should be brought about by slow degrees.

Prevention.—A successful system of prevention can only be instituted when we duly appreciate the fact that almost all cases of this intestinal fever are due to contagion. And this is precisely what our hog breeders fail to realize. No man in his senses will affect to deny that the disease is contagious, but the natural tendency is to seek for other causes in the great majority of cases. As in the case of all contagious affections that have attained a wide prevalence, this presents a number of outbreaks which can not be traced to contagion from any diseased stock, and these are at once assumed to be spontaneous, and the cause of the disease is sought in the peculiar treatment of the herd, and future prevention is attempted by the avoidance of these peculiarities. In illustration, I may quote from a letter of Mr. I. F. Hatch, an intelligent Illinois farmer, and former student of Cornell University:

In former years hog cholera has been local with us except when it first appeared, some ten or twelve years since. Then, as now, it was general, and swept everything. But since then it has been confined to a few farms or localities. Sometimes it appeared on a single farm, or perhaps on several farms, 1, 2, or 3 miles apart, all others escaping. I have a neighbor who has had it every second year since its first appearance, losing more or less hogs each time, but his is the only case of which I have been informed where it has been so regular and often.

This irregularity and local appearance deluded us into the belief that we were preventing the disease by extra care and attention, and that salt, sulphur, and ashes were a preventive; but we have been effectually undeceived this time, for hogs that have been doctored thus fared no better than those that had not.

There is one man here whose hogs have escaped the disease entirely, and he has fed for a number of years once a week, or once in two weeks, corn boiled in the ear with ashes—lyed corn, as they call it—putting a peck of wood-ashes into a 40-gallon kettle. He tells me he has had no cholera since he adopted this plan, and his hogs are certainly good subjects for cholera—poor, half-fed, and sleeping in a pile under the barn. He says others have tried this plan and successfully warded off the disease.

He goes on to quote instances of alleged prevention by feeding house-slops without corn, and by giving once a week a feed of the boiled jowls and waste parts of the pigs killed the year previous, and adds:

Generally diseased hogs run, sleep, and eat with the others, it being the general opinion that they'll all have it anyway, so no matter. I am of a different opinion. A few change the yards and sleeping places, but generally they stay in the same places throughout the disease. No attention is paid to disinfection.

These alleged preventives are doubtless somewhat beneficial by maintaining a free action of the bowels and kidneys, and favoring the elimination of the poison, as does diarrhea in the milder cases of the disease. But there need be as little doubt that, like the salt, sulphur, and ashes, they would all fail in the presence of a strong dose of the poison. Meanwhile, they are made to serve an evil purpose in diverting attention from the one effectual means of restricting the disease, the extinction of the poison. It must be fully recognized that neither contact nor proximity is necessary to contagion. The poison may be carried a certain distance on a stream of water without losing its vitality. It may be blown a long way by a favorable wind, when dried up, on light objects. It may be carried on the boots, hands, etc., of men (dealers) passing from farm to farm and from district to district. Horses, cattle, sheep, dogs, fowls, pigeons, and wild animals of all kinds are liable to carry the virus on their feet and limbs, and to deal out death to the pigs at places widely separated from each other. It is, therefore, quite impossible to trace all new outbreaks to contagion. But to attribute them to spontaneous evolution of the disease is to beg the whole question.

It can be freely conceded that a certain number of cases probably originate spontaneously every year; but these are few and far between as compared with the enormous mortality caused by contagion. It can be equally conceded that certain seasons are far more favorable to the propagation and virulence of the disease than others, yet even in these the great majority of cases are infections. It can be admitted even that a wholesome laxative diet is to some extent protective, as well as comfortable dwellings and antiseptic agents, like copperas, bisulphite of soda, charcoal, or carbolic acid. But all such protectives are comparatively limited in their operation, and, though they seem to have saved a few isolated herds, will fail disastrously if generally relied on. The epizootic influence, too, though apparently all-powerful in localities where the poison has already penetrated, fails to produce the disease in the neighboring States not previously infected. We can not give too much attention to secure the best sanitary conditions of life for the hogs, but if we allow a few of these so to engross our attention that our eyes are blinded to the most important of all—the prevention of contagion—we shall only spread the poison and increase the destruction of our herds. On the other hand, the highest success must attend such measures as will stop the production of the poison and destroy and render innocuous what is already in existence.

Diseased pigs must be removed from the healthy, killed, and buried. A thorough disinfection of all buildings, yards, and manure must be made. Chloride of lime or zinc, sulphate of iron, or carbolic acid may be used for all solids, floors, troughs, walls, etc., and for drains: and sulphurous acid or chlorine for the atmosphere. The sulphurous acid may be produced by burning sulphur, and the chlorine by adding oil of vitrol to common salt and a little black oxide of manganese. The surviving pigs must be carefully watched for the first signs of illness. Any unusual sluggishness, stiffness, or inappetence, or any disposition to leave the herd, demands a careful examination; and if there is heat or shivering, and, above all, if the thermometer introduced into the rectum indicates a temperature above 103° Fahrenheit, the animal should be at once separated from the herd, and destroyed as soon as unequivocal symptoms of the malady are shown. Care should be taken to avoid the possibility of contamination by water which has passed infected hog-pens or fields. If the malady exists within a wide radius, the visits of dealers and others must be absolutely forbidden, and a similar prohibition should attach as far as possible to quadrupeds and birds, wild and tame. Disinfectants may even be given to the sound animals that have been exposed to contagion. A table-spoonful of charcoal, animal or vegetable, may be given daily to each pig in its food. Twenty grains of bisulphite of soda, or ten drops of carbolic acid, or 10 grains of sulphate of iron may be used instead, and a tea-spoonful each of sulphur and gentian may be added with advantage. When a herd has been freed from the disease, a most exhaustive disinfection of the whole premises, manure, and other products is imperative, and it is usually desirable to change the site of the hog-pen and run to obviate any future effects of this most virulent contagion. Old and rotten wood work should be burned.

In purchasing pigs, buyers will consult their interests by avoiding markets, and going rather to the breeders whose stock is known to be healthy, and by seeing personally to the thorough cleansing and disinfection of loading-banks, cars, boats, etc., which they must use in bringing them home. And after all such precautions, newly purchased swine should invariably be placed in quarantine, at a safe distance from other hogs, and kept there for three weeks, with separate attendants, until they have been proved sound.

As in the case of other fatal contagions, this could doubtless be kept in check, or even completely extinguished, by a uniform system of destruction of the infected, and disinfection of their carcasses and all with which they have come in contact. Such a proceeding would imply an amount of governmental supervision and pecuniary outlay that would be profitable in the long run, though the past experience of the American people have scarcely prepared them to sanction it.

In 1877, a paper written by H. J. Detmers, V. S., giving his observations of the disease generally called hog cholera, and his conclusions as to its nature, was reprinted in the annual report of the Department of Agriculture. Dr. Detmers proposed to call the malady "epizootic influenza of swine," and he divided it into (1) "the catarrhal rheumatic form," (2) "the gastric-rheumatic form," (3) "the cerebro-rheumatic form," and (4) "the lymphatic-rheumatic form." The causes of the disease he divided into three classes, and he writes:

As belonging to the first class I look upon everything that will interrupt or disturb the perspiration; in the second class I place all such noxious influences and agencies as interfere directly with the process of respiration; finally, in the third class, I put all such noxious agencies or injurious influences as tend to aggravate the disease if already existing, by aiding in making its character more typhoid, or which produce a special predisposition by weakening the constitution of the animal.

In regard to the contagiousness of the affection he says:

Still I think the epizootic character of the fearful spreading of the disease can be satisfactorily explained without the existence of a contagion.

The above extracts show the most divergent views among physcians and veterinarians as to the cause and nature of the epizootics among swine. At the same time there were equally wide differences of opinion among swine-breeders and the contributors to the agricultural press. In short, there was the greatest uncertainty and confusion of ideas not only as to the characters of the disease or diseases, but as to the most elementary principles to be applied for prevention.

It was just at this period (1878) that Congress provided for the first investigation of swine diseases by making an appropriation of $10,000 for this purpose. The beginning of this investigation, which has been continued with short interruptions until the present time, marks a new era in our knowledge of swine epizootics and of contagious diseases in general. It was just at the time when the biological researches of Pasteur in regard to fermentation were attracting so much attention, and had already inspired Lister to make his discoveries in antiseptic surgery, and Koch to institute those researches which definitely connected the *bacillus anthracis* with the causation of charbon. The time was propitious, therefore, for the institution of a new line of researches, in a direction which even then promised much, and which since has practically revolutionized the position of medical science as to the nature of contagion and the methods most important for its control.

The writings of Drs. Sutton, Snow, and Law gave a very correct statement of the symptoms and *post-mortem* appearances of the organs usually found in the swine epizootics of this country. They covered about all the points which can be decided by ordinary field observations, but they left many questions still unsolved which it was necessary to determine before measures of prevention could be formulated and made successful.

There were many who believed that these epizootics were due to the ravages of more than one disease, and there were also many who held that contagion and infection played no part in their extension. As to the nature of the agent or agents which caused the outbreaks, or the conditions which were required to prevent their destructive ravages, no results of value had been obtained, or indeed could be, without a more systematic and persistent investigation, with all the instruments, apparatus, and laboratory facilities which are required to work out the obscure problems of pathological science. As with other contagious diseases of men and animals, the instruments of the *post-mortem* case had been the means by which a certain amount of information had been gained, but there were problems which they could not reveal, and for these the microscope, the culture apparatus, the biological and chemical laboratories were necessary, and without these a reliable solution could not be reached.

For the investigations of 1878 nine men were appointed for a period of two months each, as follows: Dr. H. J. Detmers, Illinois; Dr. James Law, New York; Dr. D. W. Voyles, Indiana; Dr. D. E. Salmon, North Carolina; Dr. Albert Dunlap, Iowa; Dr. R. F. Dyer, Illinois; Dr. A. S. Payne, Virginia; Dr. J. N. McNutt, Missouri; Dr. C. M. Hines, Kansas. The researches were to be made in the field in different sections of the country, and various remedies, suggested as applicable to the treatment of epizootic diseases, were to be tried. The result of this investigation may be summed up practically as follows: (1) Swine diseases were found destructive in the most widely separated districts of the country. (2) The symptoms and lesions enumerated were similar to those formerly given by Drs. Sutton, Law, and Snow. (3) No evidence was discovered to show the existence of more than one disease which prevailed as an epizootic. (4) There was a preponderance of opinion that the outbreaks were due to contagion and that the disease was communicable. (5) The remedies tested either produced no effect or were of doubtful value for the treatment of affected animals or for guarding against the contagion.

There were two points of more than usual interest raised in the investigation. In the course of Dr. Law's investigations he made inoculation experiments with rabbits, rats, and sheep, which he thought demonstrated the communicability of the disease that he investigated. Dr. Detmers devoted more of his time to microscopical investigations, and asserted that he had discovered a specific micro-organism, which he called the *Bacillus suis*, and that he had demonstrated its pathogenic connection with the disease.

Both of these conclusions must now be looked upon as premature and not supported by the direct and positive evidence which is necessary to establish such important points. That Dr. Law produced disease in the experimental animals which he inoculated is not to be doubted for a moment, but that it was the same disease and caused by the same microbe which produced the swine epizootics of the country could only be definitely determined after the microbes of the swine diseases had been identified and their characters established. Very frequently in inoculating with the products of disease taken from dead animals, septic disorders are caused in the inoculated animals which are entirely different in nature from the malady which caused the death of the individual inoculated from. To determine, therefore, whether the disease induced is identical with the disease which we propose to propagate by inoculation it becomes necessary to determine the microbe which causes the morbid changes in each case. At that time neither of the microbes which we have since identified and studied in swine epizootics in this country had been discovered, and consequently the line of evidence just indicated was impossible.

Dr. Detmers's culture experiments were too crude and primitive to be of any value, even in the condition in which bacteriological science was at that time; and his descriptions of the germ might be applied with equal accuracy to many different species of micro-organisms. For this reason no attempt will be made in this volume to give the details of either the culture or inoculation experiments which he made. While failing to obtain the evidence necessary to establish the connection of any micro-organism with the production of this malady, this and subsequent reports of Dr. Detmers had some effect in directing the attention of investigators in this country to bacteriological researches as a promising field in which to search for the hidden mysteries connected with this class of diseases.*

* It is due to justice to state in this connection that, whereas the appointments limited the work of those selected for the investigations of 1878 to a period of two months, Drs. Detmers and Law were subsequently given an opportunity to continue their investigations for a much longer period. It is also proper to state that several months before the investigations of 1878 began, Dr. Klein, of London, had published investigations which he believed established the fact that the swine epizootics of Great Britain were caused by a bacillus. See Quarterly Journal of Microscopical Science, April, 1878.

The investigations of the Department of Agriculture were continued in 1879 by Drs. Law and Detmers; in 1880 by Drs. Law, Detmers, and Salmon, and in 1881-'82 by Drs. Detmers and Salmon. The efforts of Dr. Law were largely directed to obtaining a modified virus for preventive inoculation, a doubly difficult and discouraging task before the microbes of the diseases had been discovered and studied. While his results were more or less encouraging, no conclusions of practical value as to a method of preventive inoculation were established.

Dr. Detmers continued his microscopic observations and modified his first descriptions of the microbe so far as to speak of it in later reports as a micrococcus. The author devoted the greater part of his time during these years to an investigation of southern cattle fever and fowl cholera, but incidentally investigated a number of outbreaks of swine diseases. He found micrococci in the liquids of the affected animals, which he cultivated and inoculated into other animals, but his results were not sufficiently positive to demonstrate their connection with the disease.

There was here a period of about four years when the investigations of swine diseases gave no very decided advance in our knowledge of the subject. The reason for this was that the investigation had been carried about as far as was possible by the methods then employed. To get a better insight into the nature of the epizootics and the peculiarities of the contagion, it was essential that the pathogenic microbes should be discovered and accurately studied. The investigators saw this and were working in the proper direction, but at that time bacteriological science was young and the methods of research had not been clearly worked out. Another and equal difficulty was the fact that the investigators were working without proper laboratory facilities and apparatus for such researches, without assistants, and some of them were devoting much of their time to other duties. Under such conditions it is next to impossible to reach successful results in such a difficult field of study. These difficulties were not fully appreciated either by the officials of the Department of Agriculture or by the stock-owners of the country, for the reason that such investigations were new to them and their requirements but imperfectly understood. It was not expected that a chemist would go into a sorghum field and discover the proportion of cane sugar in the sorghum cane without apparatus or laboratory facilities, but it was expected that the veterinarian would make much more difficult and delicate investigations than these with no other aids than an ax, a butcher knife, a scalpel, and a microscope.

The author saw that more facilities must be provided for these investigations, and the work systematized and properly divided, or the final result would be disappointment and failure. It was through his urgent representations that a beginning was made in 1883, by securing an unoccupied room under the roof of the Department building. which was fitted up as a laboratory, and by obtaining permission to rent a small place in the suburbs of Washington on which to keep experimental animals.

The year 1883, however, was one in which the necessity of investigating the prevalence of the lung plague or pleuro-pneumonia of cattle became too urgent to admit of delay. There were frequent outbreaks which it was the duty of the chief of the veterinary division to visit and give assistance to the State authorities as to their diagnosis and eradication. This precluded any systematic investigations of swine diseases. The spring of 1884 brought the extensive outbreaks of ergotism in Kansas, Missouri, and Illinois, which alarmed the whole country, and which demanded instant attention; and these unfortunately were followed, in the summer, by the outbreaks of pleuro-pneumonia in Ohio, Illinois, and Kentucky, which required the constant attention of the chief veterinarian and prevented scientific researches for several months.

The investigations during the years 1883 and 1884 were, therefore, too much interrupted to allow the number of experiments and the thorough working out of details which was desirable. The time was not entirely barren of results, however, in the investigation of swine diseases. The laboratory was fitted up and suitable microscopic and bacteriological apparatus obtained. Able assistants were selected and placed in charge of the different branches of the work. The lines of research were marked out and such arrangements made as would permit of the intelligent prosecution of the work, even in the absence of the chief of the bureau.

This systematization of the investigations, with the employment of a proper number of persons to keep the different lines of study advancing uniformly, was followed almost immediately by the most important developments. In 1885 a motile bacillus was discovered to be the cause of hog cholera, and its leading characteristics were accurately described. In 1886 the production of immunity by chemical products was demonstrated, the effect of disinfectants on the hog cholera contagion was thoroughly studied, and the presence of swine plague, a second epizootic disease of swine, was discovered. These discoveries solve the most important problems connected with swine epizootics and give a secure scientific basis from which to work in future. In fact these diseases are now much better understood than are most of the contagious diseases of people, and the measures applicable, in the present condition of science, to their prevention are equally as apparent as are those which are applied to the plagues of mankind.

In the sections of this work which follow the details of the investigations of hog cholera made since 1884 are given as fully as are considered necessary to an understanding of the different phases of the subject. The account of the experiments in regard to swine plague are reserved for publication in a separate volume, which it is hoped will appear within a few months.

SYMPTOMS AND POST-MORTEM APPEARANCES IN HOG CHOLERA.

The symptoms are not entirely characteristic, owing to the many forms which the disease may assume. It is moreover quite troublesome to make careful thermometric and other clinical observations on swine, which enhances the difficulty of exactly determining the course of the malady. In general we may regard the disease as manifesting itself in an acute and a chronic or mild form.

Of the acute form very little need be said. The animals die very suddenly, without having been sick for more than a few hours to a day. The chronic disease may last from three weeks to several months. The animals grow stupid and dull, they refuse to eat, and are apt to separate themselves from the rest of the herd. They grow weak, and their movements become slightly tottering. A common symptom is diarrhea, which may appear very soon after the animal becomes sick and last until it dies. In protracted cases the liver may become affected secondarily through the impaired condition of the large intestine.

The following symptoms, as noted for three or four years by Dr. Kilborne at the experiment station of the Bureau, will show how little there is upon which we can depend for a diagnosis during life.

The temperature of healthy pigs ranges between 101° and 104½° F. With sick animals it may rise from 1° to 3° above the temperature observed in health. Frequently this elevation is absent. During an outbreak elevation of temperature may be considered diagnostic, but absence of such elevation proves nothing, since an animal may die without having shown any rise of temperature during the disease. There is rarely any cold or shivering stage. The sick animals are dumpish, spiritless, and lie quietly in a corner or huddle together, hiding the head under the litter. They refuse to move even when disturbed, and are more or less oblivious to surroundings.

There is more or less loss of appetite. They usually continue to eat a little, however, and often the appetite is scarcely impaired during the whole course of the disease. Acute cases may be seen feeding before they are found dead an hour or so later. In most cases the stomach is well filled after death.

The bowels may be costive at the outset or continue apparently normal for some time, or they soon become costive, and remain so in some cases throughout the disease. In the later stages diarrhea of a liquid, fetid character appears in many cases. The color of the discharges depends largely upon the feed. In slop or garbage fed pigs they are usually of a dirty black color. In those fed with grain they are light yellowish. The diarrhea persists until death. Vomiting is absent.

Respiration is only occasionally quickened and labored in the later stages. The pulse is more rapid than in health, but its determination unsatisfactory, owing to the struggle of the animals when held. Cough is infrequent, and then only heard when the animals are aroused, as a single effort or in paroxysms.

The skin is frequently found reddened on the nose, abdomen, inside of limbs, the ears, and over the pubic region. The redness is diffuse, varying from a slight blush to a deep bluish red or purple. It increases as death approaches, and is usually found at the autopsy. A skin eruption appears occasionally, which is followed by crusts or scabs of varying size. Reddening of the skin is a sympton common to the other swine diseases—*rouget* and swine plague.

The eyes are apt to be watery. This is frequently the first outward sign. Later the discharge becomes thick, yellowish, often gumming the lids together.

Towards the end of the disease the animals have a gaunt appearance, with arched back and staring coat. Emaciation is very rapid. The weakness manifests itself in a staggering, uncertain gait, as if the animal were about to fall. Death ensues quietly. Rarely convulsive kicking is observed. The mortality is very high, usually from 80 to 90 per cent. of those attacked die.

These symptoms vary in intensity, and only a certain number are seen in one animal at the same time. In very chronic cases only the autopsy enabled us to make a diagnosis.

LESIONS PRODUCED BY HOG CHOLERA.—POST MORTEM APPEARANCES.

(See Plates I–VIII inclusive.)

a. *The acute type.*—This might with equal propriety be called the hemorrhagic type, inasmuch as the chief and perhaps the only changes are hemorrhagic in character, and these lesions are seen most clearly when an animal is examined immediately after death. The spleen is variably enlarged, soft, gorged with blood. Sometimes it is twice as long as the normal spleen (the other dimensions being proportionately increased), and it may extend across the median line to the right side. Next to the spleen the lymphatic glands and serous membranes are most severely involved. The cortex shows, on section, as a hemorrhagic line or band, according to the amount of extravasated blood, or the entire gland may be infiltrated with it. Among the glands most commonly hemorrhagic are those of the meso-colon (large intestine), those at the root of the lungs (bronchial), and on the posterior thoracic aorta. Besides these. the retro-peritoneal and gastric glands (lesser omentum) may be involved. Most rarely the mesentric glands show extravasations to a slight extent. Hemorrhages are also quite frequent beneath the serous surfaces of abdomen and thorax. They are most abundant under the serosa of the large and small intestines as petec-chiæ and larger patches. They are occasionally found under the peri-toneum near the kidneys, the diaphragm, the costal pleura as extrava-sations nearly an inch across.

The lungs, in a small percentage of cases. show subpleural ecchy-moses in large numbers. and on section small hemorrhagic foci are ob-served throughout the lung tissue. In a few cases severe hemorrhages, involving one or more lobes. have been observed. The kidneys are oc-casionally the seat of extensive hemorrhagic changes. The glomeruli appear as blood red points ; larger extravasa'ions occur in the medul-lary substance. and blood may collect around the apices of the papil-læ. The subcutaneous tissue over the ventral surface of the body may be dotted with petecchiæ and occasionally collections of blood hæmato-mata) are found in the superficial muscular tissue. The brain and spinal cord have not been examined for want of time. In one case, however, in which they were laid bare. petecchiæ were seen on the cere-bellum.

The digestive tract usually is the seat of extensive lesions. The fun-dus of the stomach is as a rule deeply reddened ; there may be more or less hemorrhage on the surface. giving rise to patches or larger sheets of blood clots. The small intestine in some cases has submucous ecchy-moses throughout its entire length. In the large intestine the ecchy-moses may be so numerous as to give the membrane a dark red ap-pearance. The food is now and then found incased in sheets of blood clot due to hemorrhage on the surface.

b. The chronic form is perhaps the most common, at least in those epizootics which we have studied. The acute hemorrhagic cases usually die in the beginning of an outbreak and are apt to be overlooked. Fol-lowing these are the more protracted. milder cases. In these animals the disease is apt to be limited in its severe manifestations to the large intestine. The other organs. however. are not exempt from degen-erative changes, owing in part to the impairment of the functions of the large intestine. consequent fermentations and the absorption of poisons thereby produced, in part to the presence of the specific bacteria in the spleen and presumably in other organs where they exercise their poison-ous activity.

The lesions of the large intestines are necrotic and ulcerative in char-acter. The ulcers may be isolated and appear as circular, slightly-pro-jecting masses stained yellowish or blackish or both in alternate rings. Or they may be slightly depressed and somewhat ragged in outline.

When the superficial slough is scraped away many ulcers show a grayish or white bottom. A vertical section reveals a rather firm neoplastic growth, extending usually to the inner muscular coat. When sections of such an ulcer are stained with aniline dyes and examined under the microscope we notice the submucous tissue very much thickened, infiltrated with round cells, and containing a large number of dilated vessels. Resting upon this thickened submucosa is a line of very deeply-stained amorphous matter, and upon this is situated the necrotic mass which fails to retain the coloring matter and is penetrated by an immense number of bacteria of various kinds. Frequently ova of *trichocephalus* are imbedded in the slough.

The extent of the submucous infiltration depends upon the age of the ulcer. In old ulcers it contains many newly-formed capillaries, and evidences of the formation of connective tissue are present. The capillaries may extend to the very edge of the border where the slough begins. The latter may have been partly shed, leaving a smooth line bounding the cicatricial tissue. The submucous infiltration gradually disappears as we leave the center of the ulcer, and somewhat outside of the ulcer no inflammation of the membrane is observable. Giant cells may be seen in some cases in the intertubular tissue at the edge of the ulcer. In very young ulcers it is frequently possible to observe the fundi of the tubules intact, while the inner or free half is converted into an amorphous mass. The depth to which the infiltration extends is not always limited to the submucosa; it may extend into the muscular coats and cause inflammatory thickening there and inflammation and the formation of new vessels in the subjacent serosa.

In some cases the necrosis, instead of appearing in circumscribed ulcers from one-sixteenth to one-half inch or more across, involves the whole surface of the mucosa, giving it the appearance of a so-called diphtheritic membrane. In such cases the walls of the intestine are very much thickened and so friable as to be easily torn with the forceps in handling. Such necrosis is rare in spontaneous cases, but it quite invariably appears in animals which have been fed with pure cultures of hog cholera bacilli.

The distribution of the ulcers varies but slightly. They appear most frequently in the cæcum and on the ileo-cæcal valve, as well as in the upper half of the colon. The lower half is implicated in severe cases only, and then less extensively. The rectum is only very rarely ulcerated. The lower portion of the *ileum* is ulcerated in a small percentage of animals, especially when they have been fed with hog cholera viscera or cultures. The stomach is occasionally the seat of slight ulceration. The lymphatic glands of the affected intestine are usually much enlarged, pale, tough, whitish on section. The spleen is rarely enlarged; the liver shows degenerative changes (softening of the parenchyma, increase of connective tissue). Heart and lungs are usually normal. The lobular pneumonia frequently found in young pigs in the winter months must be ascribed primarily to exposure rather than to the disease, as will be shown subsequently.

In some outbreaks the acute and the chronic types of the disease are not so clearly distinct as given in the foregoing pages. Frequently recent hemorrhagic lesions seem to be grafted upon cases presenting extensive ulcerations, which certainly are much older than the extravasations. It may be that the latter are the result of a secondary invasion of the hog cholera virus, either from the ulcers in the intestine or from without.

HISTORY AND AUTOPSY NOTES OF AN OUTBREAK OF HOG CHOLERA.

There is perhaps no better way to illustrate this disease than by giving briefly the history of an outbreak in a single isolated herd. In the following pages are recorded the autopsy notes and the bacteriological examination of such an outbreak near the city of Washington during the months of November and December, 1887. A study of this outbreak was taken up to ascertain more especially the condition of the lungs in hog cholera. As it is the fifth or sixth which has been observed during the past three years less attention was paid to careful bacteriological observations. At the same time this phase was by no means neglected, as the notes will show. The history of the outbreak as far as could be ascertained was briefly as follows: On October 28, there were in all one hundred and nineteen swine, chiefly young pigs, weighing from 50 to 100 pounds. Most of these had been purchased in the city markets. At this same time some twenty boar pigs were castrated. Within two weeks these began to die, and soon after the others took sick, dying at the rate of three to four a day. Less than three weeks after the first deaths only sixty-seven remained out of the one hundred and nineteen. At the end of the year only about a dozen were alive out of the entire herd. These may have acquired immunity.

The animals were kept in pens on the top of a hillock sheltered from the weather by large boxes. They were swill-fed, and this may account for their feeble resistance to the disease. In most of them there was a cirrhosis of the liver, with softening of the parenchyma, which was probably induced by the feeding. The origin of the epizootic could not be traced, as the animals had come from various quarters. The city markets had proved themselves in the past the source of disease in several purchases of pigs for experimental purposes.

The autopsy and bacteriological notes will be given in the order in which the animals were examined, any general remarks being reserved for the end. The rapidity with which the animals succumbed to the disease made any very thorough examination quite impossible, since it seemed more desirable to utilize all the material by examining those organs most always diseased in hog cholera. Most attention was paid to the lungs and the digestive tract, while the lungs and the spleen were the only organs examined bacteriologically. These notes will serve to supplement the autopsy notes in the reports for 1885 and 1886.

In making cultures from the spleen the following method was usually adopted. At the autopsy the abdomen was carefully laid open by first removing the skin and then cutting through the abdominal muscles with flamed instruments. The flaps laid back, brought into view the spleen not touched as yet by any instrument. It was then drawn out with flamed forceps, severed from its attachments with flamed scissors and placed in a large bottle plugged with cotton wool which had been previously subjected to a temperature of 150°–160° C. for two hours. In this way it was taken to the laboratory and either immediately examined or kept in the refrigerator below 55° F. over night. In making cultures the spleen was placed on a sterile glass support and the surface thoroughly charred with a red-hot platinum spatula. This was always done, although seemingly unnecessary when we consider the momentary exposure to the air in transferring the spleen from the abdomen to the sterile bottle. It may, however, destroy any bacteria which have entered the peritoneal cavity through ulcers. Through this charred area an incision or rent was made and a platinum wire introduced, and then a tube of gelatine or beef infusion inoculated with it. When roll cultures were made a minute bit of spleen pulp was torn away from beneath the charred portion and stirred about in the liquefied gelatine. From this usually a second tube was prepared. Experience of past years had shown that frequently this is not sufficient to insure the fertility of the cultures. In chronic cases with the spleen but moderately enlarged, hog cholera bacteria are found in very small numbers. In such cases bits of spleen are cut out from the charred area with flamed scissors and transferred to tubes of gelatine or beef infusion with or without peptone. Such cultures rarely fail. It might be supposed that the chances of accidental contamination are very great in this process. But a long experience with spleens of healthy animals and with organs in the study of other diseases has demonstrated the entire safety of this procedure. Salmon culture tubes with bits of organs in the bottom covered by nutrient liquids have remained sterile for months in the laboratory. At present the Esmarch tube or roll culture is indispensable in such cases.

In nearly all the cases examined both liquid and gelatine cultures were made. The former permit a diagnosis on the following day, while the latter require at least two days, usually three or four, before a reliable diagnosis can be made. The cultures were always examined unstained in a hanging drop, as the bacteria in this way are not deprived of their power of motility, which is one of the important diagnostic characters. Staining cultures was also resorted to, but it adds little information to that gained by a careful examination of the hanging drop. When gelatine cultures were examined the bacteria were always mixed with some sterile beef infusion to bring out their motility.

In a number of cases rabbits were inoculated directly from lung tissue. A small bit, about one half centimeter cube, was torn up with flamed forceps in a flamed watch glass containing some sterile beef infusion, and the turbid fluid injected beneath the skin of the thigh. The syringe used was an ordinary hypodermic syringe carefully disinfected by 5 per cent. carbolic acid above and below the piston for one-half hour both after and before use, and each time thoroughly rinsed in boiling water. As hog cholera bacteria are destroyed by a 1 per cent. solution of carbolic acid in less than ten minutes, and by a momentary contact with water near the boiling point the disinfection was certainly all that could be desired. This method was regarded as less open to criticism than the insertion of bits of tissue under the skin. We still stand in need of a syringe which can be disinfected without much trouble, as the above method is extremely tedious. The syringes devised by Koch are both unsatisfactory. The joints formed by the glass barrel and the metal cap in the syringe in which the propelling force is air were found to leak in five out of six samples. From the fluid injected into rabbits either plate or roll cultures were made in order to get an idea of the approximate number and the kind of organisms present. In every case the portion of lung tissue from which the inoculations were made was transferred to sterilized bottles and protected from accidental contamination as carefully as possible. Unless otherwise indicated, the methods just given were employed throughout the investigation.

November 16.—Pig No. 1, just died and brought to experimental station. No skin lesions; heart and lungs normal with exception of a few collapsed lobules in principal lobe of one lung.* In abdomen omentum injected so as to appear bright red. Spleen enlarged, soft, dark. Some lymphatics have the cortex hemorrhagic. Stomach and small intestines normal. In cæcum near valve several large superficial yellowish ulcers and a number of smaller ones, an eighth of an inch in diameter. Two bits of spleen were cut out and dropped into a tube of gelatine, and one of beef infusion. The gelatine culture remains sterile. The liquid culture contains the motile hog cholera bacteria and a large butyric bacillus growing only in the bottom of the liquid.

No. 2 died last night. Buzzards have eaten into the thorax and penetrated the coats of the stomach. Only the small ventral lobe of right lung diseased. Bright red, mottled with pale yellow dots. (See Plate IX, fig. 2.) The smallest bronchi occluded by cylindrical plugs. Alveoli likewise occluded. The plugs consist chiefly of cells and are so dry and firm that they may be removed as small ramifications or branchings when the lung tissue is torn away. Bronchial glands enlarged, hemorrhagic. Spleen enormously enlarged, liver in advanced stage of cirrhosis. Glands at portal fissure chronically enlarged. In cæcum four superficial ulcers one-half inch across, slough stained yellow. In upper colon four similar to these and a large number of small ones about one-eighth inch across with yellowish slough.

A moderate number of bacteria found in cover-glass preparations of spleen. A gelatine tube culture contained, after three days about, seventy-five to one hundred colonies of the motile hog cholera bacteria in

* In order to understand the description of the lung lesions, the following brief outline of the anatomy of the lung and of the terms used may be of service:
The right lung is made up of four lobes: the left has only three. (In text-books on anatomy the left lung is considered as being made up of only two.)
In both there is a large principal lobe resting upon the diaphragm and against the adjacent thoracic wall. This lobe forms the major part of each lung. The remainder, occupying the anterior (or cephalic) portion of the cavity, is made up of two small lobes, one extending ventrally (or downward in the standing position of the animal) and in the expanded state covering the heart laterally, the other extending towards the head and overlapping the base of the heart. These small lobes may be denominated the ventral and cephalic lobes, respectively. The right cephalic lobe is longer and more distinct from the ventral lobe than the corresponding left cephalic. Wedged in between the two principal lobes and resting on the diaphragm is a small lobe, pyramidal, belonging to the right lung (azygos lobe). This lobe rests on the left against the mediastinal membrane, and on the right it is separated from the right principal lobe by a fold of the pleura passing from the ventral abdominal wall to inclose the inferior vena cava. This small lobe is almost completely shut off, therefore, from the other lobes by folds of the pleura.

each needle track. A liquid culture inoculated simply with the platinum wire thrust into the spleen pulp contained hog cholera bacteria only. A rabbit was inoculated from lung tissue as above described; about one-fourth of a cubic centimeter of the suspension injected. Dead on the seventh day. Slight fibrinous exudate on coils of intestines. Spleen very large, soft, dark. Beginning coagulation-necrosis in liver. Bacteria very numerous in spleen; both gelatine and liquid cultures contain only the motile hog cholera bacteria. The plate from lung tissue with which rabbit was inoculated contains a very large number of identical colonies, made up of motile hog cholera bacteria.

November 17.—No. 3, male, died yesterday. Buzzards have consumed pectoral muscles and pierced into thorax and abdomen. Ecchymoses on costal pleura and entire epicardium, a few under pulmonary pleura. Left lung hypostatic, slight amount of whitish foam in trachea. Lung tissue otherwise entirely normal, neither hepatization nor collapse anywhere to be seen. Bronchial glands and those along posterior aorta with hemorrhagic cortex. Ecchymoses in the subcutaneous fatty tissue over entire ventral aspect of body, about one-eighth inch across, beneath peritoneum of abdominal muscles and of the entire length of small intestine, from size of pin's head to one-eighth inch. Congestion in patches in large intestine, no ulceration. Stomach and intestines contain a yellow liquid resembling the yolk of eggs. Spleen but slightly enlarged. Ecchymoses under serosa of liver. Cirrhosis moderate. A bit of spleen tissue dropped into a tube of beef infusion gave rise to a culture of hog cholera bacteria. In each needle track of a gelatine culture countless colonies of the motile bacteria appear.

November 18.—No. 4, medium sized male, died this morning. Superficial inguinal glands enlarged, cortex slightly hemorrhagic. Some fibrils on coils of intestine. Petecchiæ under serosa of small intestine; extensive hemorrhage between mucous and muscular coat of stomach along fundus, forming a clot about one-half inch thick. Spleen very large, friable, blackish, extends beyond median line into right side. Petecchiæ on liver, which is considerably cirrhosed; mucosa of stomach along fundus blackish; closely set petecchiæ under mucosa of small intestine; contents liquid, blood-stained. A dark hemorrhagic patch near valve in large intestine. Colon studded with petecchiæ, no ulceration. Small number of petecchiæ on surface and throughout cortex of kidneys. No bacteria seen on a cover-glass preparation of spleen tissue. Beef infusion inoculated with a wire thrust into spleen remains sterile. A bit of spleen dropped into a tube of gelatine gives rise to about a dozen colonies of hog cholera bacteria.

November 18.—No. 5, male, 75 pounds, three to four months old, died last night. Inguinal glands enlarged, pale. Left cephalic, ventral, and about one-third of principal, right cephalic and ventral lobes of lungs solidified, bright red, mottled with minute yellowish dots, *i. e.*, same as No. 2. Costal pleura covered with a very thin whitish exudate. Solidified portion of principal lobe adherent. On epicardium a very delicate papery deposit. Spleen much congested. Inflammatory adhesion of liver to diaphragm. Mesenteric glands nearly as large as hen's eggs, mottled, pale red. About twelve ulcers one-quarter to one-half inch across in lower ileum; base depressed and covered with a thin yellow layer; outline irregular; not connected with Peyer's patches. Valve thickened and ulcerated. Several ulcers in cæcum; a large projecting slough attached to one of them. A gelatine and a liquid culture from the spleen contain hog cholera bacteria only. A liquid culture from the pleura remains sterile.

A rabbit inoculated from the hepatized lung tissue died on the eighth day. Slight fibrinous exudate on intestines, diffuse coagulation necrosis in liver, spleen much congested. Immense number of hog cholera bacteria in spleen and liver. Cultures pure. A plate made from the same bit of (pig's) lung tissue contains about fifty colonies resembling hog cholera.

No. 6, large black male, died last night. Extensive sero-fibrinous infiltration of subcutis from axilla to pubis and over right thigh. The subcutis has a gelatinous aspect. Spleen much congested. Lungs normal. In cover-glass preparations from spleen large bacilli, with ends square. Cultures in gelatine and beef infusion remain sterile. Animal probably died of malignant œdema.

November 21.—No. 7, medium sized male, died yesterday. Temperature of air varying from 26° to 50° F. Omentum deeply reddened. Spleen enlarged, soft; small hemorrhages under capsule one-eighth inch across. Liver imparts a sensation of grittiness when cut. Cortex of kidneys dotted with hemorrhagic points. The patch of mucous crypts about valve in cæcum dark, pigmented; no ulceration. Mucosa of stomach hemorrhagic in fundus. Thorax half full of blood-stained serum and some fibrin stretching from lung surface to walls; small quantity of fibrin in pericardial cavity. Lung tissue infiltrated with blood (pulmonary hemorrhage). Two ventral lobes collapsed; subpleural hemorrhagic patches. A cover-glass preparation from spleen contains numerous hog cholera bacteria. In each needle track of a gelatine culture, countless colonies. A liquid culture from a bit of spleen tissue contains hog cholera bacteria only.

No. 8, small female, died yesterday. Patches of skin on ventral aspect of limbs and groin reddened. Minute petecchiæ under serosa of ventral abdominal walls. Spleen very large, congested. Some delicate fibrils stretched over coils of intestine. Liver cirrhosed. Glands of mesentery and meso-colon enlarged; cortex hemorrhagic. Fundus of stomach moderately congested. In cæcum are black pigment spots resembling former hemorrhage, chiefly on the summit of folds. In upper colon ulcers about three to a square inch, each one-fourth inch in diameter, and covered by a convex, projecting yellowish slough. Ulcers found down to rectum. Small quantity of clear serum and a few fibrils in pleural cavities and pericardium. Left ventral and tip of right cephalic lobe collapsed. A gelatine and a liquid culture made from a bit of spleen tissue contain hog cholera bacteria and a butyric bacillus which slowly liquefies the gelatine.

No. 9, small male. Diffuse reddening of ventral aspect of body especially marked on limbs. Spleen, lymphatics, and kidneys normal. A small abscess in pelvis attached to bladder, probably caused by castration. Venous congestion of vessels of meso-colon. The transverse folds of mucosa of colon covered by very thin yellowish patches of necrosis. Feces hard, distending the large intestine. Cephalic and ventral lobes of left lung solid, plainly mottled. The bronchioles and alveoli filled with consistent plugs of cellular exudate. Remainder of lungs normal. Owing to small size of spleen, two liquid cultures were made, each with a bit of spleen tissue. In both, hog cholera bacteria alone appeared. A rabbit inoculated with hepatized lung tissue remains well for weeks after. A plate culture therefrom is liquefied in two days.

No. 10, small castrated male, died yesterday. Has been sick for sometime according to overseer of farm. Spleen not enlarged. Inflammatory changes in pelvis due to castration. Stomach slightly reddened and bile stained. One large ulcer on ileo cæcal valve one-half inch across, black, with yellowish margin. A few very superficial ulcers in cæcum. Lungs normal, left somewhat hypostatic. Two cultures made with bits of spleen contain both hog cholera bacteria and butyric bacilli.

No. 11, dying, killed by being bled from brachial vessels. Lungs normal. Spleen small. Liver slightly cirrhosed. Two ulcers in cæcum; slight superficial necrosis in colon. A liquid culture made from a bit of spleen contains hog cholera bacteria on following day.

November 23.—No. 12, small white male, died yesterday. Median line of ventral aspect of body much reddened, limbs slightly so; extravasation under skin and into muscular tissue over sternum. Much blood-stained serum in peritoneal cavity; spleen very large, dark. In kidneys, cortex and base of pyramids deeply reddened. In cæcum, slight traces of superficial necrosis. Anterior and cephalic lobes of both lungs collapsed. Blood-stained serum in pericardium. Numerous hog cholera bacteria on cover-glass preparations of spleen pulp. In a gelatine culture from the spleen countless colonies appear in each needle track. A liquid culture made with platinum wire contains hog cholera bacteria on following day. A rabbit inoculated from the collapsed lung tissue died on the tenth day. Spleen engorged; numerous foci of coagulation necrosis in liver, involving each one or more acini; extensive necrosis along border of left lobe. Ecchymosis of pyloric valve and duodenum; hemorrhagic foci in lungs, about fifteen in each lung, one eighth inch to three-sixteenths inch in diameter. Examination of the spleen shows numerous hog-cholera bacteria. A gelatine culture contains countless colonies in each needle track.

Number 13, small female, died yesterday. Skin on ventral aspect of limbs and over pubis reddened. Spleen greatly enlarged, congested. Lungs normal. A few lobules in ventral and right cephalic lobe collapsed. Liver slightly cirrhosed. Stomach distended with food. Large patch of mucosa in fundus reddened. Cæcum and colon contain numerous ragged depressed ulcerations. Valve entirely encircled by ulceration. Contents of intestine liquid, yellow. A gelatine and a liquid culture inoculated with a platinum wire thrust into spleen remain sterile. No bacteria seen in cover-glass preparations.

No. 14, small female, weight about 50 pounds. Considerable reddening of the skin over ventral aspect of body and limbs; especially marked along median line. Superficial inguinals enlarged, of a mottled pale and deep red on section. Spleen very large, 12 inches long, 2 inches broad, and five-eighths inch thick at hilus; gorged with blood, friable. A small number of punctiform hemorrhages in cortical portion of kidneys. Glands of mesentery and of colon enlarged and congested. Deep reddening of several square inches of mucosa in fundus of stomach. Large intestine contains a semi-liquid mass, chiefly earth. Four large ulcers in cæcum, one of them at least 1 inch across, covered by a yellowish slough; the peritoneum covering it is thickened and inflamed. In upper colon there is considerable necrosis, involving the epithelium in patches. Lungs normal, excepting the right ventral lobe, which is solid. Bronchi and air cells of this lobe completely occluded by plugs as with No. 2; surface bright red, mottled with yellowish points—the ultimate air cells filled with the cellular exudate. Subpleural ecchymoses over both lungs. From the spleen a liquid and a gelatine culture contained only hog-cholera bacteria. They were very numerous in cover-glass preparations from this organ.

A rabbit, inoculated from the consolidated lung tissue died on the seventh day. At the point of inoculation a pasty mass extends to abdomen, only subcutis involved. Spleen engorged. Single acini and groups in the liver are completely necrosed, yellowish white. In both organs, hog-cholera bacteria. Cultures from spleen pure.

November 25.—No. 15. black and white male, died yesterday morning. Redness of skin of abdomen, throat, and limbs. Superficial inguinals hemorrhagic. Spleen very large. gorged with blood. Lungs normal, excepting a few lobules at the caudal border of principal lobes, which are red, collapsed, and contain lung worms. Pericardium contains deep-colored serum and coagula; left auricle dotted with petecchiæ. Bronchial lymphatics hemorrhagic. Liver slightly cirrhosed. Cyst in right kidney one-half inch in diameter. Fundus of stomach intensely congested, similarly the cæcum and colon; no ulcers. Lymphatics of meso-colon hemorrhagic. A beef-infusion culture from a bit of spleen contains the motile hog-cholera bacteria only. A gelatine culture became liquefied by the heat of the laboratory; no bacteria seen on a cover-glass preparation of spleen.

No. 16, black and white female, died last night. Skin and spleen as in previous case. The small ventral lobe of both lungs collapsed, lungs otherwise normal. Liver in advanced stage of sclerosis; stasis of portal circulation. Lymphatics of abdomen as in No. 15. Fundus of stomach slightly reddened; one ulcer three-fourths inch across. Extensive necrosis of mucous membrane in colon and rectum, slight in cæcum; wherever the membrane is free from a slough it is deeply congested. Numerous hog-cholera bacteria and some large (butyric) bacilli in spleen. A beef infusion and a gelatine culture contain them. From a bit of collapsed lung tissue a plate culture is made and a rabbit inoculated. The plate contains in two days about six to seven liquefying and a large number of non liquefying colonies, the latter made up of motile hog-cholera bacteria; the rabbit died on the seventh day. Spleen very large, friable : contains large numbers of hog cholera bacteria. On left lobe of liver an area of necrosis one-fourth inch by three-fourths and one-sixteenth inch deep. On the right lobe only three or four acini necrosed. Pylorus and duodenum covered with hemorrhagic dots and patches. Culture from spleen pure.

November 26.—No. 17, female of medium size; died suddenly this morning without previous illness. Subcutaneous fat abundant. Spleen moderately congested. Petecchiæ in cortical portion of kidneys. Fundus of stomach slightly reddened. In lower ileum patches of congestion. Scattered petecchiæ in mucosa of large intestine; contents normal. Glands of mesentery and meso-colon with cortex hemorrhagic. Large quantity of blood-stained serum in thorax. Fibrinous deposit on pleura; lungs partly expanded ; cephalic half of right lung solid, blackish; air-tubes and alveoli filled with extravasated blood. Interlobular tissue distended with blood-stained serum (pulmonary hemorrhage). Left lung in the same condition. Trachea full of reddish foam. Blood and fibrinous coagula in pericardial cavity. Beef infusion, into which a bit of spleen was placed. remains sterile; also a tube of gelatine inoculated from the spleen with platinum needle. In a cover-glass preparation of the spleen large (butyric?) bacilli.

No. 18, small female, died last night. Buzzards have consumed thigh muscles. Diffuse reddening along median line of abdomen. Hemorrhagic spots one-eighth to one-fourth inch across, subepidermal, chiefly on ventral aspect of limbs; subcutaneous and subperitoneal ecchymoses. All abdominal lymphatics with cortex infiltrated with blood. Serosa of large intestines as if sprinkled with fresh blood; several subserous hemorrhagic spots one-half inch across on diaphragm. along inferior vena cava, common bile duct, and gall bladder; also, under mucosa of the whole length of small intestine, throughout cortical portion and between pelvis and medullary portion of kidneys. Fundus of stomach one mass of petecchiæ and larger extravasations. One ulcer in cæcum, old, with indurated base. Valve and the patch of mucous crypts at its base ulcerated; depth of ulcer indurated, consisting of a tough, pale tissue. Small old ulcers in upper colon. Lungs dotted with small hemorrhages, chiefly subpleural. Three or four hemorrhagic patches under costal pleura of each side. Lung tissue normal, excepting the base of right ventral and the tip of left ventral lobe, which are collapsed. Hemorrhages under epicardium over entire heart; left auricle one mass of ecchymoses. Coagula of fibrin in auriculo-ventricular groove.

Through an oversight the cultures from this animal and the one following were both numbered the same, so that it was impossible to identify them. A liquid culture from each was made by adding a bit of spleen tissue. A gelatine culture from each was made simply with platinum wire. One tube of infusion contains hog cholera and butyric bacteria; one tube of gelatine contains immense numbers of hog-cholera colonies. The other two tubes remain sterile. A rabbit inoculated from the collapsed lung tissue remained well for a month after. A plate culture from the same contained about six colonies, evidently of hog-cholera bacteria.

No. 19, small male, died last night. Diffuse reddening on abdomen along median line. Superficial inguinals very large, pale, œdematous; peritonitis; feeble adhesion of coils of intestine to ventral wall and of lobes of liver to one another. Slight fibrinous deposit on intestines. Liver sclerosed. Lymphatics hemorrhagic. Cortex of kidneys dotted with extravasations. Mucosa of large intestines of a dark slate color; it is dotted with closely-set conical elevations, tough, whitish, about one-eighth inch high and one-sixth across. When scraped away a depressed pale pink, sharply-outlined spot remains. Microscopic examination of the intestine showed that these elevations correspond to amorphous masses, which failed to become colored on applying the ordinary staining agents. They covered portions of the mucosa which were either wholly or partially necrosed and incapable of being stained. In some places the outline of the tubules could still be discerned. On applying Weigert's fibrin stain, long meshes of fibrin corresponding in general to the outline and position of the destroyed tubules appeared. Valve thickened and completely covered by ulceration. Cephalic half of both lungs airless, affected with broncho-pneumonia; catarrhal exudate filled alveoli and air tubes. Epicardium dotted with hemorrhagic points. Slight, feeble, pleural adhesions.

For cultures from spleen, see No. 18. A rabbit inoculated with lung tissue died on the fifth day. Spleen slightly enlarged; contains many hog-cholera bacteria. Gelatine culture contains very many colonies of the same.

No. 20, a large female. No skin lesions; a few strings of coagula over coils of intestine; spleen small; base of pyramids of kidneys much reddened; ileum dotted with subperitoneal hemorrhagic points and patches; mucosa not affected; mesenteric glands with cortex hemorrhagic. Large intestine empty, scattered ulcers one-eighth to one-fourth inch across. Liver sclerosed.

A beef-infusion culture inoculated with a bit of spleen tissue contains hog-cholera bacteria and butyric bacilli. A gelatine tube culture contains a few hog-cholera colonies.

November 29.—No. 21, small Jersey red, female, died last night. Skin deeply reddened along median line of abdomen. Large quantity of blood-stained serum in peritoneal cavity. Serosa of ileum and the mesentery completely covered with hemorrhagic points and patches. Lymphatics of abdomen hemorrhagic throughout their substance. Petecchiae in cortex of kidneys. Mucosa in fundus of stomach hemorrhagic. Mucosa of lower ileum a confluent layer of necrosed tissue. Numerous round ulcers in caecum and colon: slough projects slightly. Lungs hypostatic, interlobular spaces distended with coagulated blood, most marked in dependent lobes. Simple collapse of both ventral lobes and of the right principal near the root. Ventricles of heart dotted with petecchiae, auricles black, covered with clotted extravasated blood. Hog cholera bacteria present in cover-glass preparations from spleen and in both gelatine and liquid cultures.

No. 22, male, Jersey red, died last night. Glands of abdomen with cortex infiltrated with blood. Spleen engorged. A few hemorrhagic patches in fundus of stomach. In caecum and colon ulcers about three-eighths inch across; very thin, adherent, yellowish slough. About one ulcer to 4 square inches of surface. Considerable blood-stained serum in pleural cavities, hemorrhage in anterior half of left lung, which is solid, blackish on section. Clots and reddish foam in trachea and bronchi; hemorrhage seems somewhat older than in No. 21. Pleura of right lung roughened and parts adherent to chest wall. Ventral lobe solid, bronchioles firmly plugged with dry catarrhal secretion. Hog cholera bacteria on cover-glass from spleen pulp and in a gelatine and liquid culture made therefrom; only five to ten colonies in each track of the platinum wire. A rabbit was inoculated from a bit of the solidified lung tissue and a plate culture made. The latter develops numerous colonies of hog cholera bacteria. The rabbit died on the thirteenth day. Numerous hog cholera bacteria in both organs. A gelatine culture from the spleen contains the same organisms.

No. 23, small black male. Buzzards had removed pectorals of one side. Lymphatics congested, not hemorrhagic. Crowded petecchiae in subcutis and beneath peritoneum of ventral abdominal wall. Spleen very large, congested. Four or five small hemorrhages on diaphragm. Mucosa of caecum and upper colon ulcerated in large patches. In lower colon the ulcers are small, yellowish, depressed areas embedded in a very dark mucosa dotted with numerous punctiform ecchymoses. At root of left lung there is some collapse, extending slightly into all lobes. Of right lung the ventral and cephalic lobes are collapsed, with occasional emphysematous lobules interspersed. Lungs otherwise normal. Extensive hemorrhage of both auricles in the form of diffuse patches and petecchiae. A gelatine and a liquid culture from the spleen remain sterile. A plate culture made from a bit of collapsed lung tissue contains a considerable number of colonies, which are made up of bacteria resembling those of hog cholera, but differing from them in their manner of growth on gelatine, in beef infusion, and in the absence of motility. A rabbit inoculated from the same bit of lung tissue remains well. A plate culture from a bit of normal tissue from the same lung contains but one colony.

No. 24 died yesterday; medium-sized female. Spleen very large, extends beyond median line. Lungs normal, excepting collapse of a small portion of the right ventral lobe. Lymphatic glands in general with hemorrhagic cortex. Caecum and colon very much congested, ulceration superficial and slight. Entire fundus of stomach of a uniform deep wine color. A gelatine and a liquid culture from the spleen contain only motile hog cholera bacteria. In the former the colonies are

very numerous. The bacteria are demonstrated in cover-glass preparations from spleen pulp.

No. 25, medium sized black male, died yesterday. Spleen very large.
Subperitoneal tissue full of petecchiæ. Lymphatics with hemorrhagic
cortex. Hemorrhages under serosa of duodenum and common bile duct.
Valve completely ulcerated. Intestine otherwise normal and feces
dry. Lungs contain a large number of subpleural hemorrhages, tissue
normal, epicardium hemorrhagic. Kidneys contain a specimen of kid-
ney-worm (*sclerostoma pinguicola*). A gelatine and a beef infusion culture
from the spleen became confused with those of another pig (No. 435), to
be described later on. One set of cultures remained sterile; the other
contained hog cholera bacteria. In all probability the sterile cultures
belonged to this animal.

December 2.—No. 26, medium-sized female, died last night. Considerable redness of skin over the ventral aspect of limbs and along median line of abdomen. Spleen very large, blackish. Several strong
fibrous adhesions betweeen costal and pulmonary pleura. Lung tissue
normal. Left kidney contains a small cyst. Large number of round
depressed ulcers stained yellow, most numerous in cæcum. One ulcer
is one-half inch across, the inflammation extending to serosa. Stomach
very dark, pigmented along fundus. On a cover-glass preparation
from spleen numerous large bacilli (probably butyric). They did not
develop in the liquid and gelatine culture, which latter contained only
hog cholera bactera colonies, very numerous in each needle track.

No. 27, small black male. Superficial inguinals enlarged, pale, œdematous. Superficial ulceration and hemorrhagic changes in cæcum.
Lungs normal. *Post mortem* changes too advanced for cultures.

No. 28, small female, died last night. Superficial inguinals with
cortex hemorrhagic. Spleen slightly enlarged. Minute hæmatomata
on its surface. Slight sclerosis of liver. Left kidney contains six cysts
one-half inch diameter. Stomach in fundus somewhat congested. In
large intestine, ulcers one-quarter to three-eighths inch diameter, with
adherent yellowish, projecting slough, most numerous in cæcum. About
one-third of the ventral lobe of each lung and a few lobules of the right
principal lobe collapsed. Lungs otherwise normal. In a beef infusion
culture from the spleen both hog cholera and butyric bacteria were
present. A gelatine culture contained but two colonies. From a bit of
collapsed lung tissue a plate culture was made and a rabbit inoculated.
On the plate about fifteen colonies of hog cholera bacteria appeared.
The rabbit died on the fifteenth day. Bare indications of necrosis in
liver. Spleen very large. Both organs contain a moderate number of
hog cholera bacteria. A roll culture from the spleen is pure.

No. 29, large black male, died about thirty-six hours ago. Temperature below freezing point. Skin deeply reddened over entire ventral
aspect of body. Superficial inguinals enlarged, slightly congested.
Spleen very large (about 14 inches long), very soft and friable. Slight
sclerosis of liver and old perihepatitis. Retro-peritoneal glands with
cortex hemorrhagic. Cyst in right kidney. Slight ecchymoses in fundus of stomach. Valve and patch of mucous glands in cæcum very dark
with pigment. Slight superficial ulceration in cæcum and upper colon.
Lungs and heart normal. Both gelatine and beef infusion cultures contain hog cholera bacteria. In the former the colonies are very numerous.

No. 30, small black male; died thirty-six hours ago. No reddening
of skin. Superficial inguinals and spleen but slightly enlarged. Peritonitis. Petecchiæ under serosa of small and large intestines and blad-

der. Moderate amount of fibrinous exudate. Valve completely ulcerated, ulcer deep. One near valve has caused thickening of serosa of intestinal wall. In cæcum an extensive patch of ulceration; in colon a few ulcers and numerous punctiform hemorrhages. Right lung completely adherent to costal pleura by means of fibrous tissue. A small ventral portion of principal lobe of left lung is airless, pale red, mottled with yellowish points. Air tubes and vesicular portion occluded with dry cylindrical plugs. A culture in gelatine from the spleen develops numerous colonies in each track of the wire. A liquid culture contains the butyric bacillus also. A bit of the solidified lung tissue was used for a plate culture and to inoculate a rabbit. The plate developed countless colonies of hog cholera bacteria as tested by other cultures. The rabbit died on the tenth day. The spleen was enlarged, the liver full of centers of advanced necrosis, involving one to three acini; hog cholera bacteria very numerous in both organs.

December 3.—No. 31, small black and white female, died yesterday. Well-marked odor of decomposition. No skin discoloration. Spleen but slightly enlarged. Lymphatics in general with hemorrhagic cortex. Liver in state of advanced sclerosis. Stomach normal. A few large old ulcers and a considerable number of small ones throughout cæcum and colon. Mucosa itself pigmented with patches of fresh congestion. Lungs normal. A liquid culture from the spleen contains only butyric bacilli.

No. 32, small black and white female. No skin discoloration. Lymphatics, including inguinal, bronchial, and peritoneal with more or less hemorrhagic cortex. Spleen engorged. Liver sclerosed. Kidneys with numerous petecchiæ throughout cortical portion. Stomach pale. Several large old ulcers in upper colon. Cæcum and colon pigmented. Lungs normal with exception of a few collapsed lobules in ventral lobes. Both a gelatine and a beef infusion culture contain hog cholera bacteria. Colonies very few.

No. 33, large black and white male; said to have died last night. Advanced *post mortem* changes. No examination made, excepting to see the condition of lungs, which were healthy.

No. 34, medium-sized white, died last night. Considerable reddening over ventral aspect of body. Inguinals reddened on section. Small quantity of dark-colored serum and numerous yellowish flaky coagula attached to abdominal organs. Spleen very large, dark, friable. Liver cirrhosed. Kidneys with cortical portion thickly dotted with petecchiæ, hemorrhage into pelvis. Lymphatics in abdomen with cortex more or less hemorrhagic. Extensive and deep ulceration in large intestines; in cæcum and upper colon as large patches, in lower colon as small ulcers. The mucosa which is not destroyed is deeply congested. Stomach along fundus deeply reddened. Extensive fibrous adhesions of left lung to walls of thorax. Collapse involves ventral, cephalic, and portions of principal lobe of right lung, and small portion of principal and ventral lobe of left lung. Extensive muco-purulent secretion in trachea, bronchi, and subdivisions throughout both lungs. Bronchial glands with cortex hemorrhagic. A beef infusion culture from the spleen contains hog cholera and butyric bacteria. In the gelatine culture each track of the wire contains countless colonies of what are shown under the microscope to be motile hog cholera bacteria. From a bit of collapsed lung tissue a plate culture was made and a rabbit inoculated. The plate develops countless non-liquefying colonies. The rabbit died on the eighth day. Spleen enlarged. Liver infested with coccidia. A slight amount of coagulation-necrosis. Hog cholera bacteria in both organs and in cultures from the spleen (both gelatine and liquid).

December 5.—No. 35, small black and white female, died yesterday. Redness of skin over abdomen and inside of limbs. Spleen enlarged, slightly congested. Lymphatics with cortex infiltrated with blood. Large quantity of blood-stained serum in abdomen. Liver sclerosed. A few scattered petecchiæ in cortical portion of kidneys. Mucosa of colon pigmented; a few small ulcers present. Lungs normal, with exception of a few collapsed lobules in ventral lobe of each lung. Dark-colored serum in pericardial cavity. The spleen contains large numbers of butyric bacilli. A liquid culture contains both hog cholera and butyric bacilli. A gelatine culture remains sterile.

No. 36, large black and white female, died yesterday. Lungs normal. Spleen very large, dark, friable. Liver sclerosed; lymphatics generally with cortex congested or hemorrhagic. Large number of ulcers in cæcum and colon; mucosa deeply congested. From spleen countless colonies of hog cholera bacteria in a gelatine tube culture. In a liquid culture the butyric bacilli are also found.

No. 37, medium Jersey red and the last of a lot of seven, dead forty-eight hours. This animal has been sick for some time; the spleen was very large, gorged with blood. Lymphatics pale. Large number of old ulcers, from one sixteenth to 1½ inches across in cæcum and colon; mucous membrane generally pale. Of the lungs, both ventral and a small portion of cephalic lobes with smaller bronchi and alveoli plugged with dry catarrhal products. A liquid culture from the spleen contains hog cholera and butyric bacilli. A gelatine culture contains countless colonies. The spleen pulp on coverglass preparations shows many. From the diseased lung tissue a rabbit was inoculated and a plate culture made. The latter develops a large number of colonies of hog cholera bacteria. The rabbit died on the eighth day. Spleen enlarged and friable, contains many hog-cholera bacteria. Slight coagulation-necrosis in liver, which is infested with coccidia. Cultures from blood and spleen pure.

December 7.—No. 38, small white female, died last night. Much emaciated. Subcutaneous and subperitoneal tissue contains numerous extravasations. Lymphatics in general with hemorrhagic cortex. Spleen moderately congested. Lung tissue normal. Numerous subpleural and epicardial hemorrhages. Kidneys with cortical portion dotted with numerous punctiform extravasations. Mucosa of small intestine contains numerous petecchiæ. One ulcer, one-half inch across, in lower ileum. Extensive and deep ulceration throughout whole length of large intestine. Some of the ulcers over 2 inches across; surface coal-black; the inflammation extending through the intestinal walls to serosa, which is dotted with scattered extravasations. A gelatine tube culture from spleen pulp contains countless colonies in each needle track, consisting, as seen under the miscroscope, of motile hog cholera bacteria. A liquid culture contains also streptococci and butyric bacilli.

No. 39, black and white female, *post mortem* changes under way. No examination made beyond ascertaining that lungs are normal, lymphatics with cortex infiltrated with blood, and spleen enlarged and congested.

December 8.—No. 40, small black female, died this morning. Superficial inguinals enlarged but pale. Moderate quantity of straw-collored serum in abdominal cavity. Spleen enlarged, deeply congested. In cæcum and upper colon a large number of deep broad ulcers; in some the inflammation extends through intestinal wall to serosa. In lower colon, ulcers small and mucosa deeply congested. Numerous hog cholera bacteria in spleen as shown by a gelatine culture. A liquid culture contains also butyric bacilli.

No. 41, small black and white female, died yesterday. *Post mortem* changes under way. Large quantity of bload-stained serum in abdomen. Plastic peritonitis matting together the various organs; spleen enlarged and congested. Numerous old ulcers in cæcum and colon, with adherent slough. Adhesive pleuritis, with large quantity of blood-stained serum in thorax. Lungs normal, excepting areas of collapse in ventral and cephalic lobes. A liquid culture from spleen contains both hog cholera and butyric bacilli. A gelatine tube culture develops a large number of colonies in each needle track.

December 10.—No. 42, small black and white male, died yesterday morning. No discoloration of skin. Spleen very large, congested. Lymphatics with cortex slightly reddened. Liver cirrhosed. A few extravasations in pyramids of kidneys. Mucosa of cæcum and upper colon covered with ulcers, the adherent slough dirty yellowish. Fully one-half the area of membrane thus involved, the remainder is pale. Lungs normal. From the spleen, hog cholera as well as butyric bacilli appeared in a beef-infusion culture. In a tube of gelatine the colonies were very numerous.

December 12.—No. 43, small black and white male, died December 10. No discoloration of skin. Spleen very large, friable, gorged with blood. Inguinals and lymphatics at lesser curvature of stomach hemorrhagic throughout. Those of mesentery and meso-colon less so. A few petecchiæ on surface of kidneys. Liver slightly cirrhosed. Mucosa of cæcum pale, of colon considerably congested. A few small ulcers with hemorrhagic border. Mucosa in fundus of stomach deeply congested. Slight hemorrhage in principal lobe of each lung, otherwise both normal. Bronchial glands hemorrhagic. Hog cholera bacteria quite abundant in spleen, as shown by cover-glass preparations and cultures.

December 15.—No. 44, medium-sized white female, died yesterday morning. Slight reddening of skin along median line of abdomen, limbs, and throat. Spleen enlarged and engorged. Inguinal, bronchial, retroperitoneal and meso-colic glands with cortex hemorrhagic. Liver badly cirrhosed. In cæcum one ulcer an inch across, involving entire thickness of wall, and few smaller ones. A few in upper colon. Lungs œdematous. At least one-half of each lung (most dependent portion) airless, of a red flesh color. Sprinkled through it in some places more densely than in others are grayish-yellow areas one-half to two millimeters in diameter. Trachea full of foam. Bronchi contain a thick mucous secretion, most abundant in the diseased region. A roll culture from the spleen contains numerous colonies of hog cholera bacteria. From a bit of lung tissue a rabbit was inoculated; a plate culture from the same shows a very large number of colonies, probably hog cholera. The rabbit died on the seventh day. Spleen enlarged. Coagulation necrosis in liver, which also contains coccidia. Hog cholera bacteria numerous in spleen, as shown by cover-glass preparations and roll cultures.

No. 45, large black-and-white female, died December 13. Slight reddening of skin. Spleen large, congested. Lymphatics in general with cortex hemorrhagic. Liver badly cirrhosed; surface dotted with hemorrhagic points. Kidneys on surface and on section, as well as mucosa of cæcum and upper colon, dotted with numerous petecchiæ. No ulceration. Large quantity of blood-stained serum and coagula in pleural sacs, chiefly in the right. Lungs not collapsed, infiltrated with a reddish serum; left hypostatic. Abundant mucous secretion in bronchi and smaller air tubes stained with blood. No hepatization. Roll culture from a bit of spleen contains a large number of colonies of hog cholera bacteria. The same may be seen in cover-glass preparations from the spleen itself.

No. 46. Large black-and-white male, died December 13. Spleen enlarged and congested. Abdominal lymphatics with cortex more or less congested. Liver slightly cirrhosed. One large ulcer on valve about 1 inch across; several half as large in the middle of colon. Mucosa deeply congested in cæcum and colon; much pale serum in pleural sacs. Lungs slightly œdematous. Of right lung the principal near root, ventral and tip of cephalic airless, collapsed. One lobule on ventral (diaphramatic) surface of principal lobe airless, with faint catarrhal injection of bronchioles and alveoli; of left lung a small portion of the principal collapsed; the ventral and cephalic emphysematous. The trachea, bronchi, and branches contain a large quantity of a translucent viscid mucus; no lung worms present. Bronchial glands enlarged, pale; those on posterior aorta with cortex hemorrhagic. Extravasations beneath epicardium. A plate culture from a bit of lung tissue develops about seventy-five colonies of hog cholera bacteria. A rabbit inoculated from the same bit remains well. A roll culture from the pig's spleen contains about fifty colonies alike; one examined is made up of motile hog cholera bacteria.

December 17.—No. 47, small female, died yesterday. Superficial inguinals entirely hemorrhagic. Mesenteric glands slightly congested. Spleen barely congested. Mucosa of lower ileum completely ulcerated. Similarly that of cæcum and upper colon, but more severely so as to make the intestinal wall very friable. In lower colon the ulcers are isolated. Lungs and heart normal. In a roll culture from a bit of spleen tissue about one hundred colonies of hog cholera bacteria appeared.

No. 48, medium-sized black-and-white female, died last night. Superficial inguinals enlarged, pale. Abdominal lymphatics in general with cortex hemorrhagic. Spleen congested; surface covered with numerous elevated blood-red points. Liver extensively cirrhosed. Mucosa of ileum dotted with petecchiæ. It contains about fifteen ulcers, not limited to Peyer's patches, with longer diameter transverse, in some cases encircling the tube. The ulcer is covered by a thin yellow slough. One ulcer, one-half inch across, in cæcum, and two in colon. About one-half pint of straw-colored serum and a mass of semi-gelatinous pale coagulum the size of a fist in each pleural sac. Lungs but partly collapsed, pleura slightly roughened. Interlobular tissue of dependent lobes distended with serum, parenchyma œdematous, so as to sink in water. Bronchi contain a slight amount of reddish fluid. In a roll culture from a bit of spleen pulp about two hundred colonies of hog cholera bacteria appear.

December 24.—No. 49, large, black and white. Spleen gorged with blood. Lymphatics generally pale; liver cirrhosed. Cæcum and colon with walls thickened and very friable; mucosa entirely ulcerated. One ulcer in rectum. A roll culture from the spleen melted, but found to contain on examination only hog cholera bacteria.

DISEASE IN HEALTHY PIGS CAUSED BY MATERIAL FROM THIS EPI-
ZOOTIC.

November 21.—Two pigs (Nos. 434, 435), about two months old, fed with spleens from several of the preceding cases. Two days later a few more spleens were given them. Both became sick a few days later and died November 28, about twelve hours apart.

No. 434. Spleen slightly enlarged, full of blood. All glomeruli of kidneys show as hemorrhagic points: lymphatics moderately congested. Stomach along entire fundus deeply congested. Superficial small ulcers in cæcum and colon; in the cæcum they are covered by a projecting slough. Right ventral and cephalic lobes collapsed, the former developed into broncho pneumonia (catarrhal injection of small air tubes and alveoli). A few lobules of left principal lobe in the same condition. Hog cholera bacteria in the spleen as determined by a gelatine and a liquid culture. A rabbit was inoculated with a bit of lung tissue and a plate culture made. This developed countless colonies, non-liquefying, alike, shown to be motile hog cholera bacteria. Rabbit dies on sixth day. Spleen moderately congested; contains many hog cholera bacteria. In liver, minute foci of necrosis. A gelatine culture from the spleen contains numerous colonies of hog cholera bacteria.

No. 435. Spleen enlarged, covered with hemorrhagic elevations. Kidneys hemorrhagic as in No. 434. Stomach near pylorus deeply congested. Cæcum and entire colon covered with a dirty yellowish and blackish slough. Right and left ventral lobes, a small portion of right principal and left cephalic affected with broncho-pneumonia. Cultures from the spleen of this animal were confused with those of another pig (No. 25), but one set remained sterile; the other contained hog cholera bacteria, and it is highly probable that the fertile cultures belonged to this animal.

A few additional cases are cited to show the infectious nature of this outbreak.

No. 436 and No. 437 were placed, November 27, in the infected pen containing the two preceding animals. They were also fed portions of hog cholera viscera later on. No. 436 was found dead December 27. Red blotches on skin of ventral aspect of body. Superficial inguinals hemorrhagic. Other lymphatics enlarged but pale. Spleen slightly congested. Mucosa of cæcum and colon deeply congested and dotted with considerable number of small ulcers. A roll culture from the spleen melted, but contains only hog cholera bacteria according to microscopic examination. No. 437 did not take the disease.

No. 449 and No. 452, placed in the same pen December 17, but not fed with infectious matter, died December 29. The lesions were somewhat different from those usually found, and are briefly as follows:

No. 452. Spleen and lymphatics not enlarged. Large quantity of serum and fibrinous coagula in abdomen. Viscera generally agglutinated. Lungs glued to chest wall by a recent exudate. Pericardium distended with serum and coagula. Lung tissue not affected. Kidneys deeply reddened. Mucosa of cæcum and colon entirely covered with a thin layer of diphtheritic exudate; when scraped away a deeply reddened surface is exposed. Numerous small, deep ulcers present. A roll culture of spleen, also melted from the heat of laboratory, contains only hog cholera bacteria. In No. 449 the lesions were the same, excepting the pericarditis. The ulceration of large intestine less extensive.

Two very instructive cases of hog cholera were caused by simply exposing pigs on an infected asphalt floor in a pen adjoining cases of the disease.

No. 464 and No. 466, about three and one-half months old, exposed with six others January 4. No. 464 died January 11. Superficial inguinals

normal. Those in abdomen much tumefied and hemorrhagic throughout. Spleen enlarged, friable, with hemorrhagic points. Several patches
of mucosa in fundus of stomach one-fourth to one-half inch across,
covered with blood clots. In large intestine only a few scattered petecchiæ on mucous membrane. Some subpleural hemorrhages in lungs;
lung tissue normal. Scattered petecchiæ on epicardium of auricles and
ventricles. On cover glass preparations from spleen pulp a moderate
number of bacteria were present. In several roll cultures only colonies
of hog cholera bacteria appeared.

No. 466 died January 13. Skin of ears, throat, nose, limbs, and belly
deeply reddened. Spleen as in No. 464. A few petecchiæ on epicardium.
Kidneys as in No. 464. Urine contains blood. The mucosa of large intestine in general deeply congested and studded with about fifty ulcers
one-fourth inch across. Meso-colic and retro-peritoneal lymphatics
with cortex hemorrhagic. Stomach as in No. 464. Lungs normal, excepting collapse of two-thirds of ventral lobes. Roll cultures from a
bit of spleen pulp gave the same result as in preceding case.

BRIEF SUMMARY OF THE IMPORTANT FEATURES OF THIS EPIZOOTIC.

The high percentage of mortality in epizootics of hog cholera like
the foregoing is the first thing to claim our attention. Out of 119 animals not less than 100 perished within the brief space of two months,
or over 80 per cent. As no disinfection was resorted to, no isolation of
the healthy attempted, it is difficult to say what number could have
been saved. At any rate, the above figures indicate the mortality of
this disease when left to itself, and it shows that nearly all young
animals, such as weigh between 50 and 100 pounds, are susceptible to
this disease.

Most of the animals died rather unexpectedly. Only a comparatively
small number were visibly diseased some time before death. Since in
many there was more or less ulceration in the large intestine, it indicates that animals may be in a very bad condition and become a source
of infection for others without necessarily showing it.

The swill feeding has already been mentioned as a probable cause of
the cirrhosis of the liver observed in so many of these animals. This
organ was tough and imparted a gritty sensation to the hand when cut.
The parenchyma was softened and degenerated. It seems reasonable
to suppose that this chronic malady may have made the herd more
susceptible to the disease.

Hemorrhagic lesions.—At least one-third of the cases examined showed
lesions of a hemorrhagic character. The most common was an infiltration of the cortical portion of lymphatic glands with blood; sometimes
the entire gland appeared hemorrhagic on section. As regards the relative frequency of this condition, the bronchial, posterior mediastinal
(aortic), and inguinal glands stood first; next, the retro-peritoneal,
meso-colic glands and those in the lesser curvature of the stomach.
The mesenteric glands were rarely affected. Accompanying this condition of the lymphatics was a very large spleen, its great size being
simply due to an engorgement with blood.

Next in frequency were the hemorrhagic lesions of serous membranes, in the form of punctiform extravasations, larger ecchymoses, and very rarely of collections of blood infiltrating the muscular layers beneath the serous membrane. These extravasations were most frequent on the auricles and ventricles of the heart, under the serosa of the large and small intestines. Beneath the pulmonary pleura, and in the subcutaneous tissue. In the severest cases blotches appeared on the diaphragm and costal pleura. In about 10 per cent. the kidneys were hemorrhagic. The glomeruli then appeared as minute blood-red points. To this may be added hemorrhages in the pyramids and extravasations collecting around the papillæ.

The mucous membrane of the stomach in hemorrhagic cases was deeply reddened in the fundus; in some cases there was hemorrhage into the membrane, more rarely on the surface. The mucosa of the small intestine was usually intact, but that of the large intestine in the acute form of the disease was in the same condition as the stomach. In older cases, when not covered with ulcers, it was either pigmented or dark red, chronically congested. This outbreak was characterized by hemorrhagic lesions more than any other which we have examined. Our experience has been that the early cases are hemorrhagic and are succeeded by those in which ulceration, cellular infiltration of the lymphatics and signs of weakness, such as serous effusions, predominate.

In some of the animals in this outbreak there were most extensive hemorrhages. In one the mucous membrane of the stomach was separated from the muscular coat by an extensive clot one-half inch thick. In five cases (10 per cent.) the lungs were the seat of extensive hemorrhages, which literally converted the most dependent lobes into a blood clot and filled the pleural sacs with blood-stained serum. In a considerable number both peritoneal and thoracic cavities contained much blood-stained serum.

Ulcerative lesions.—Ulcers of the large intestine were present in 36 out of 49 cases, or 70 per cent. They varied from very slight to very severe and extensive lesions, involving in a small number nearly the whole mucous membrane of the cæcum and colon. The rectum was quite invariably free from disease. The age of the ulcers can not be determined, as the process of necrosis and subsequent ulceration seems to vary much in rapidity. In a few cases it was not limited to the mucous membrane, but extended into the muscular wall, producing considerable local inflammation and thickening of the serous membrane. In rare cases the necrosis and cellular infiltration had made the intestinal wall so friable that it broke when handled. When the ulceration was slight it was frequently confined to the ileo-cæcal valve and adjacent membrane, which consists of a large patch of lymph follicles and some mucous glands. The ulceration in this situation was accompanied by an extensive neoplastic thickening of the valve beneath the ulcer, indicating that the ulcer was old. In 5 cases (10 per cent.) the lower ileum was ulcerated (Nos. 5, 21, 38, 47, 48); the ulcers seemed to have no relation to Peyer's patches.

Complications.—Peritonitis, pleuritis, and pericarditis were not uncommon complications usually accompanying old ulceration.

Lung lesions.—This epizootic was studied mainly for the purpose of determining the condition of the lungs in hog cholera. Swine plague is essentially a disease of the lungs, secondarily of the digestive tract. It may be possible to find some cases of swine plague in which the large intestines are primarily diseased. Thus far they have not come to our notice. From the facts obtained from this epizootic we may safely assume that hog cholera produces no lesions which may not be found in the lungs of apparently healthy animals of the same age, and which may be due to the debility caused by the infectious disease. We must accept, however, the hemorrhages found in a small percentage of cases. Such are co-existent with hemorrhages in most other organs, and are not specific lung lesions. (See Nos. 7, 17, 21, 22, 43, of autopsy notes.) The lesions found on *post-mortem* examination were either simple collapse or lobular broncho pneumonia following it.

Simple collapse usually involved the two ventral dependent lobes, more rarely portions of the small cephalic and the principal lobes. The collapsed lobes, or groups of lobules interspersed among emphysematous lobules, appeared slightly, if at all, depressed. The color approached that of red flesh. In only a few instances could plugs be found occluding the bronchi. Sections made from lobules in this condition show a number of interesting features. The alveolar walls are crowded together in some places till they almost touch one another. Besides the fibrin, there may or may not be one or several large cells, round, with much protoplasm inclosing a vesicular nucleus. The bronchi are all open; the epithelium intact. The alveolar walls are not changed, nor is there any round cell infiltration to be seen. In circumscribed areas the capillary net-work is distended with blood corpuscles, while all the larger vessels are similarly filled with these elements. In the alveolar duct there is now and then considerable fibrillar fibrin well brought out by Weigert's stain.

In about 15 per cent. of the animals examined one of the smaller ventral lobes was airless throughout and moderately enlarged. Viewed from the surface, the diseased lobe is bright red, dotted with minute, pale grayish or yellowish points of a diffuse hazy outline, each not more than 1 millimeter ($\frac{1}{25}$ inch) in diameter. They are usually arranged in groups of four, and represent the ultimate air-tubes filled with cellular exudate. The larger bronchi are also occluded. The exudate is yellowish white, so firm that it is possible to tear away the lung tissue with needles without necessarily breaking up the inclosed exudate. It may thus be teased out in the form of branching cylinders, becoming smaller, and finally dwindling down to the size of a coarse hair.

In microscopic sections the alveolar walls are found beset with distended capillaries. The alveoli are filled up with cellular masses, fibrin appearing very rarely. In most alveoli the cells are large, round, with vesicular nucleus, evidently derived from the alveolar epithelium. In some alveoli and in the smallest air-tubes the cell mass is so dense that individual elements can only be seen with difficulty, but they appear to be identical with the cells just described. The process seems to be accompanied with very little inflammation. The desquamation and proliferation goes on in the alveoli and smallest air-tubes until they are occluded by the casts described. Of the 49 animals of the same herd, 17 were found with collapse (Nos. 1, 7, 8, 12, 13, 16, 18, 21, 23, 24, 28, 32, 34, 35, 41, 44, 46), and eight with lobular broncho-pneumonia (Nos. 2, 5, 9, 14, 19, 22, 30, 37; see also Nos. 434 and 435). More than one-half, therefore, had some defect of the lungs.

It might be questioned whether such lesions as those of broncho-pneumonia are not due to swine plague bacteria, since they closely resemble the appearance found in many swine plague lungs. This question is effectually disposed of by the inoculation of lung tissue into rabbits. From 16 lungs 16 rabbits were inoculated. Of these lungs 8 were involved in simple collapse; 8 in broncho-pneumonia. Of the 16 rabbits 4 survived (Nos. 9, 18, 23, 46); the remainder died of hog cholera (Nos. 2, 5, 12, 14, 16, 22, 28, 30, 34, 37, 44, 434). (The notes on these rabbits will be found in the autopsy notes of the swine as numbered.) Of the 4 survivors 3 had been inoculated from collapsed lung tissue, 1 from a broncho-pneumonia. It is interesting to note that of these rabbits 1 died in six days, 4 in seven days, 3 in eight days, 2 in ten days, 1 in thirteen days, and 1 in fifteen days after inoculation. Plate cultures from the corresponding bits of lung tissue showed a variable number of colonies almost invariably non-liquefying, and in many cases identified as hog cholera bacteria.

These facts lead to the inference that in hog cholera the specific bacteria will find their way to any diseased portion of lung tissue, and there multiply to a certain extent. In one case a plate culture from a bit of normal lung tissue showed but one or two colonies, while a bit of collapsed tissue from the same lung showed a large number. There is no doubt that the slight exudate and feeble circulation in collapse and the abundant partly cellular, partly mucus or fibrinous exudate into the air spaces in broncho-pneumonia furnish a favorable nidus for pathogenic bacteria. These may have been carried there by the blood or they may have been introduced from without. If the latter supposition prove true, and there are no valid objections to it, diseased lungs in hog cholera may not only become the means of disseminating the disease through the mucus and expired air, but they may become the channel, the weak spot, through which the virus enters the organism.

To elucidate this question, if possible, the following instructive experiment was made:

Two pigs (Nos. 460 and 461), about ten weeks old, received each into the right lung, December 21, 3 cubic centimeters of a beef-infusion peptone culture two days old, inoculated from a single colony growing in a roll culture. This had been made from a bit of spleen tissue from No. 46 of the outbreak described in the preceding pages. There were about fifty colonies in the tube, all alike. To test the culture, a rabbit received at the same time one-ninth cubic centimeter subcutaneously in the thigh. It died in five days. The spleen was much enlarged, blackish, friable, and contained hog cholera bacteria. A roll culture contained numerous colonies after two days. The liver contained no centers of coagulation necrosis, as the animal had succumbed too quickly.

No. 460 became very weak in its hind limbs in less than a week; respirations short and quick; bowels relaxed. It was found dead on the ninth day. Superficial inguinal glands normal. Petecchiæ in the slight deposit of fatty tissue beneath peritoneum of abdominal muscles. Spleen about 12 inches long, 1½ inches wide, and three-fourths inch thick at the hilus, blackish, friable. A few petecchiæ on cortex of left kidney. One cyst, the size of a large pea, in medulla. A large number of small hemorrhages in connective tissue around pelvis of right kidney; five small urinary cysts not showing on surface. Glands in lesser omentum enlarged, hemorrhagic throughout. In cæcum and colon an almost continuous yellow sheet of superficial necrosis, about 1 millimeter thick, covering the mucosa. In lower colon it breaks up into isolated patches, simulating ulcers. In microscopic sections this layer is found to consist of necrosed epithelium intermixed with some round cells. On Peyer's patches in lower ileum a yellow, soft deposit rests, which is not adherent and might be mistaken for chyle. Lobes of right lung glued together and to pericardium. Pleura thickened generally; serum very slight in amount, blood stained. On lobes of left lung, which are also glued together, and on right lung, there is a very slight deposit, about one-half millimeter thick, in the form of a net work. The pleuritis and exudate is most marked on the most dependent portions of the lungs. Cavity of pericardium normal. Lung tissue not hepatized anywhere; trachea and bronchi contain a small quantity of reddish fluid. Bronchial glands and those along posterior aorta hemorrhagic throughout. Cultures from pleural cavities, as well as those from spleen, contain only hog cholera bacteria. As shown in roll cultures they were very numerous in the latter organ.

While No. 460 presented such a well-marked case, No. 461, although presenting at first the same symptoms, slowly recovered. The difference may have been due to the fact that with No. 460 a 6-inch needle was used, while with No. 461 one only 3 inches long. In the latter case the chances for the passage of bacteria into the lung tissue and thence into the intestines were much poorer.

On July 31, over seven months after inoculation, No. 461 was found dead. It had continued well and thrifty, and no more attention had been paid to it. Only the lungs and the liver were brought to the lab-

oratory, since all the other organs, including the digestive tract, were
reported normal, excepting the kidneys, which were said to be highly
congested.

Lungs, but slightly collapsed, dark red. Cephalic and ventral lobes
of both lungs and the azygos lobe solid to the touch. f a grayish red
color, with tortuous injected vessels under the pleura. On section, the
tissue cuts like cold meat. Color grayish to dark red. The cut ends of
bronchi show plugs of a glairy mucus. The marginal portion of the
lobe is grayish, homogeneous, very dense. Towards the center of the
lobes the tissue is more reddish, infiltration not so dense, and speckled
with small masses (one-half millimeter across) of a yellowish-white,
homogeneous, cheesy matter. On the border of the right cephalic lobe
an encysted mass, cheesy, yellowish, gritty to the knife, evidently the
result of the inoculation. There were reported adhesions of the diseased lobes to chest wall, indicative of pleuritis.

Bronchial glands slightly larger than walnuts, of a uniform grayish
dense texture. Trachea and bronchi occluded with a blood clot of very
recent origin. In the extremities of the large bronchi lung-worms embedded in the clot.

Heart of normal size, rather flabby. Both auricles and the attached
vessels filled with tarry semi coagulated blood. In left auricle a white
clot also present. Left ventricle firmly contracted.

Liver small; surface roughened with adherent flakes of tissue, indicating old perihepatitis. Parenchyma tough, resembling in texture
very soft rubber. Acini with dark center and pale bloodless periphery.

The immediate cause of death was pulmonary hemorrhage; the re-
mote cause, the broncho-pneumonia caused by the inoculation. The
lung worms aggravated the lesions already present. What is most interesting in this connection is the fact that hog cholera bacteria were
still present in the lung tissue, as the following will show:

Three roll cultures made from bits of lung tissue; the developing
colonies all alike and resembling those of hog cholera bacteria; no
liquefying colonies in any tube. Beef infusion cultures from the individual colonies demonstrate the identity of the bacteria with those of
hog cholera. A rabbit inoculated with sterile beef infusion, in which a
bit of lung tissue had been torn up, remained alive, while another rabbit, inoculated from the pure culture made of the colonies of the roll
culture, died in six days with enlarged spleen, coagulation necrosis in
liver, extravasation in lower large intestine, and many hog cholera bacteria in spleen. Cultures from blood and spleen pure.

The lung disease in this animal could not have been more than three
months old, and was very likely due to the injury to lungs resulting
in adhesion to chest wall and the inflammation around the encysted
mass.

This experiment shows (1) that hog cholera bacteria, when introduced into the lungs, do not produce a specific parenchymatous inflammation of themselves; (2) that they may pass from the lungs by way
of the pharynx into the digestive tract and there produce their characteristic effect.

Bacteriological observations.—The preceding experiments on rabbits and the intrathoracic inoculation in case of the pigs are sufficient of themselves to establish the fact that the bacteria described in the reports of the Bureau for the years 1885 and 1886, and again found in this epizootic, are the cause of hog cholera. It may be added, however, that out of fifty-six cases (here reported) hog cholera bacteria were found in the spleen of all but six. Even in these the cultures made were too few to make the negative evidence of any value.

In many cases the hog cholera bacteria were associated with a rather large bacillus, which, for the sake of convenience, we will call butyric bacillus.* This organism was only detected when a bit of spleen was dropped into beef infusion, with or without peptone. The culture, kept at about 35° C., contained on the second and third days a cloudy mass limited to the bottom of the tube. The cloud was made up of bacilli, rather large, with a spore in one extremity of the rod strongly refracting the light. The rod was not enlarged at this end in the fresh state. When dried and stained, the shrunken protoplasm gave the spore-bearing end a swollen appearance, reminding one of the tailed bacteria of older writers. In the few tubes in which this bacillus alone was present the liquid itself remained perfectly clear; when hog cholera bacteria were present, it became uniformly but faintly clouded. In liquid cultures, without the bit of spleen, the bacilli did not develop. This was evidently necessary as food material. In gelatine-tube and roll cultures the bacilli did not grow. They are very likely anaërobic organisms, abundant in the alimentary tract, and absorbed from ulcers or lesions of blood-vessels into the circulation before death, in the spore state, and their development kept in check until that occurs. It is also probable that they are important factors in the rapid changes which may take place after death. They are quite constantly found in the liver of different animals when *post mortem* changes have begun to develop.

In some half dozen cases decomposition was so far advanced that no thorough examination was made. At first it was thought that the animals had been dead several days, but the person in charge of the herd asserted that they had died during the night. Although the temperature had fallen below 30° Fahrenheit (1° C.), decomposition was far advanced. It may be that the live animals crowded upon the dead and thus kept the bodies warm. Yet this supposition is not capable of accounting for the rapid changes. The hemorrhagic lesions may have enabled various bacteria to become distributed throughout the body. The heat disengaged by them during multiplication, aided by the body heat of the animals still alive, may have been sufficient to keep up the process of decomposition. This *post mortem* growth may also account for the large number of hog cholera bacteria found in many spleens,

* Whether this bacillus is identical with the bacillus of malignant œdema, as has been asserted by some, I do not know, as no experiments were made to test its pathogenic power.

although the temperature of the air was far below the point where multiplication may take place.

In the foregoing epizootic, as well as in those studied in 1885, and 1886, there was no difficulty whatever in demonstrating the presence of the hog cholera bacillus in the spleen. The herd referred to in the preceding pages was swill-fed, and the animals were very likely crowding each other more or less, not being compelled to hunt for food. Hence each one was exposed to a large quantity of virus. The same may be said of the penned pigs at the Experiment Station. The diseased animals found in different localities were brought to the station and penned with healthy ones. A severe epizootic was quite invariably the result, owing to the unusually good opportunity for infection in the pens and the saturation of the floors and soil of the pens with virus.

When swine roam over a considerable extent of territory in search of food, the virus is more widely distributed but less concentrated. Less virus is therefore taken up by individual animals, and although the disease is equally fatal in the end, the course may be somewhat different and the lesions less extensive. At the same time the bacteria may elude observation. They may remain more or less localized, owing to the reactive power of the organism, which destroys those that have entered the internal organs. To those who would give up the search for hog cholera bacilli after a few unsuccessful attempts to find them, we would reccommend the perusal of the following three cases, after having reviewed the epizootic just described:

Hog cholera prevailed more or less in Montgomery County, Maryland, during the latter weeks of September and the early part of October, 1888. October 17 Mr. H—— lost about 22 out of a herd of 55 to 60 swine during the past four weeks. Of those now scattered in a large field 2 appear ill; 1, a small black shoat, is killed by cutting its throat and examined. The superficial inguinal glands are very much enlarged, the surface mottled, dark red. The spleen large, but pale and rather firm. The liver shows signs of invasion of the *sclerostoma pinguicola*. The lymphatic glands at lesser curvature of stomach are very large; cortex completely hemorrhagic.

The left lung normal; the principal lobe of the right lung has in it a mass of tissue involved in broncho pneumonia, extending obliquely from the free border to near the dorsal region, about 1 inch thick; the lymphatics along the dorsal aorta are likewise hemorrhagic; the stomach filled with food; small intestines contain a number of attached *echinorhynchi*; the large intestines distended with semi-solid fecal matter; the mucosa in general is normal, but in the caecum are two ulcers about three-eighths inch across, round, slightly elevated, with center black and periphery yellow; beneath the superficial slough is a whitish, firm, new growth, extending to the muscular coat in the center of the ulcer.

The spleen and the right lung were taken to the laboratory. From the former cultures were made on *agar*, in gelatine and beef infusion, by adding bits of spleen tissue as large as peas. In no tube did any development take place. A rabbit inoculated by tearing up a piece of hepatized lung tissue in sterile beef infusion and injecting the turbid liquid subcutaneously remained well.

Several miles from the first farm we came upon a herd of young pigs which were just showing signs of disease, although none had been lost. One of them, with unsteady gait, which hid in the litter under a shed and returned to it when driven away, was killed by bleeding from the vessels of the neck. The lungs were without a sign of disease. Spleen enormously enlarged and gorged with blood. The lymphatic glands of groin and about stomach very large but rather pale, and œdematous on section. Stomach filled with food. Large intestines overdistended with very dry, hard feces, somewhat softer near cæcum ; in the latter only one ulcer and this on the valve, about one-fourth inch across and of the same nature as the one found in the preceding case.

A portion of the spleen of this animal was taken to the laboratory and cultures made, as in the previous case, with bits of spleen. All cultures remained permanently sterile.

Several miles from the latter place we found the disease on a farm situated on a hill. The swine were allowed to go a considerable distance down the slope to a marshy stream. The owner had lost 6 or 8 out of a herd of 20 to 25 within six weeks. A few were evidently ill, but none were killed, as a dead one was found. It had probably died during the night. The buzzards had consumed nearly all the intestines through a small hole near the pubis. Putrefaction had already set in. Spleen enlarged, slightly congested. In the small portion of the large intestine, which still remained, an ulcer was found three-eighths inch across. The glands of lesser omentum with hemorrhagic cortex. The stomach contains a small quantity of bile stained fluid. Both lungs glued to chest wall by coagulated fibrin from blood extravasation. Left lung contains about ten to fifteen hemorrhagic foci, visible under pleura, one-fourth to one-half inch across. The principal lobe of right lung solid, granular, involved in broncho-pneumonia. The hepatized lobe was discolored by recent and extensive blood extravasation. A gelatinous deposit under sternum resting on pericardium. The semi-decomposed condition of the animal prevented a more careful examination. Portions of the spleen and hepatized lung tissue were taken for examination.

While the spleen of the two preceding cases showed no indications of bacterial life on cover-glass preparations, the spleen of this case contained a considerable number of bacteria resembling hog cholera bacilli very closely. On gelatine they grew differently from the latter, and the cultures emitted a slightly offensive odor. In liquids they were actively motile. They were putrefactive bacteria, without effect upon two rabbits inoculated with large quantities of the cultures. A rabbit inoculated with the diseased lung tissue remained well. The latter on closer examination had a texture as granular as the roe of fishes, the granules being inspissated cell masses in the alveoli and air tubes. At least four different kinds of bacteria were present in large numbers.

The absence of bacteria from the spleens of Nos. 1 and 2 is in harmony with the results obtained in other infectious maladies when animals are killed in the early stages or during the height of the disease. It is only in the last stages that the bacteria become most numerous and appear in sufficient numbers in the internal organs to be easily detected. In the third case, death was very likely brought on by pulmonary hemorrhage not infrequently found in hog cholera. The specific bacilli produced at first the ulcers, and were either working their way slowly into the internal organs or else were being destroyed in the ulcer itself.

The latter termination would signify recovery; the former death. These ulcers might be aptly compared to the malignant pustule in man, in which the virus remains at first localized but may spread throughout the system after a time. The presence of numerous ulcers in swine in the first epizootic is to be regarded as a multiple infection, while in the three cases just cited the infection was limited to a few foci or but one. The ulcers would no doubt have revealed the virus, but our previous experience with spleens of diseased swine made it seem unnecessary to study the ulcer itself. As regards the lung disease of the third case, nothing positive can be said. It resembles most closely chronic swine plague. The germ of this disease was not present, however, as shown by the rabbit inoculation.

Buzzards may carry the disease from one place to another. When the dead animals are at all exposed to view they are immediately attacked. Whether hog cholera bacteria are entirely destroyed in the digestive tract of the birds can not be said, but there is nothing in the range of our knowledge of bacteria which will exclude the probability that the bacteria are not all destroyed during the process of digestion, and that they may be distributed by these birds from place to place in the discharges.

THE CAUSATION OR ETIOLOGY OF HOG CHOLERA.

The suspicion entertained by those engaged in the study of diseases of man and the lower animals that infectious or communicable diseases are due to living organisms of the lowest order, capable of rapid multiplication by the process of fission and spore-formation, has been transformed into conviction during the past ten years. A considerable number of the most common, most dreaded diseases have been proved to be caused by exceedingly minute, plant-like organisms known under the general name of bacteria. Among animals the micro-organisms causing anthrax, black quarter, tuberculosis, glanders, strangles, infectious pneumonia in horses and swine, and *rouget* in swine have been very thoroughly demonstrated. The opinion has been steadily gaining ground that in order to control infectious diseases we must learn their causes and the life history of the pathogenic bacterium found in each disease. These will suggest to us the measures that are most likely to prove successful in combating such maladies. Moreover, it is pretty well accepted to-day that the prevention of infectious diseases is the main thing to be arrived at in our studies, and that their treatment when they have once obtained a foothold is at best tentative and rarely successful. This is especially true of the lower animals. They cannot be treated with the same care which is accorded to human beings attacked by infectious diseases. They are (in some diseases at least) always scattering the living virus and thereby endangering those still free from disease.

In the investigation of swine epizootics these facts have been carefully borne in mind, so that most attention has been paid, first, to the life history or biology of the micro-organism; second, to disinfectants as destroyers of the specific bacteria; and third, to the various methods of preventive inoculation.

THE BACILLUS* OF HOG CHOLERA.

During the past three years the number of swine affected with this disease which have been examined is about 500. About three-fourths

---

* The term *bacterium*, implying a form genus between micrococcus and bacillus, has been almost wholly given up by bacteriologists, and all those forms classed under bacteria have been thrown together under the genus *bacillus*. This change is unfortunate for several reasons, inasmuch as many species were best classed under the genus *bacterium*. This latter term, which was applied to hog cholera microbes in the reports of 1885-'86-'87, is now reluctantly given up for the sake of uniformity.

of this number died at the experimental station of the bureau. The remainder came from outbreaks of the disease within 2 or 3 miles of the Station. Besides these, a small number of animals were examined in some of the Western States. From perhaps 400 the same microbe was obtained, there being practically no difference between the pathogenic microbes obtained from all the outbreaks thus far investigated.

Of the internal organs the spleen contains the largest number of bacteria, and in acute hemorrhagic cases they are sufficiently numerous to be detected on cover-glass preparations. A minute bit of spleen pulp is rubbed on a cover-glass, dried and heated according to the accepted methods, and then stained for a few minutes in an aqueous solution of methyl violet. The bacteria then appear as elongated ovals or short rods with rounded ends, chiefly in pairs. When the staining has been very brief only the periphery of the rod is deeply stained, the central portion being pale and simulating the appearance of an endospore. When the period of staining is prolonged to half hour or more, the rod may become uniformly stained.

Bacteria thus dried on a coverglass and mounted in balsam measure from 1.2 to 1.5, occasionally 1.8 micro-millimeters (.00005–.00006 inch) in length, and about .6 micro-millimeter (.000024 inch) in breadth.

In sections of the spleen from acute hemorrhagic cases the bacilli may be found in considerable numbers. Sections were hardened in alcohol, cut dry, and subsequently stained with aniline water methyl violet. They were in some cases decolorized in 1 per cent. acetic acid. The bacilli then appear as short, plump rods, with ends rounded off; sometimes they are short enough to deserve the name of ovals. The periphery is more deeply stained than the central body of the rod. They measure on the average 1.5 micro-millimeters. In such sections the bacteria appear in small masses in the capillary spaces of the spleen-pulp, rarely among the cells of the malpighian corpuscles (Plate X, fig. 2). The masses appear more or less star-shaped. The bacilli are crowded together in the center; from this, linear groups radiate into the capillary network. Such masses may be 8 to 10 micro-millimeters in diameter. They are fairly numerous in spleens from acute cases. That the size of these colonies is not due to *post-mortem* multiplication is shown by the fact that the largest and most numerous colonies were found in the spleen of an acute case which was examined within two hours after death, in the month of November.

In sections of ulcers hog cholera bacilli have been searched for, but the examination of a large number of ulcers showed that no positive results could be obtained. Different ulcers showed different bacteria, sometimes large colonies of micrococci, sometimes groups of large bacilli, following the course of the blood-vessels in the embryonic tissue under the slough. These no doubt found their way in from the superficial slough which seemed to be made up almost entirely of bacteria. Moreover, hog cholera bacilli closely resemble many putrefactive forms, so that even if they could be readily seen nothing but a good differential stain would enable us to recognize them. That they are present, however, may be demonstrated by inoculating mice or rabbits with bits of the ulcer. A small number of mice may succumb to malignant œdema. The rest will die of hog cholera. In rabbits the local effect of such inoculation is usually quite severe, owing to the putrefactive bacteria introduced at the same time.

Staining of hog cholera bacteria.—On cover-glass preparations they are easily brought into view by a few minutes' contact with watery solutions of the ordinary aniline colors, such as methyl violet, fuchsin and methylene blue. Decolorizing agents, such as acetic acid, one-half to 1 per cent. must be used carefully lest the color disappear from the bacilli also. When stronger dyes, such as alkaline methylene blue or aniline water fuchsin, are employed, the bacilli are decolorized with greater difficulty. Watery solutions should therefore be employed only for cover-glass preparations where decolorizing is unnecessary. For sections it is well to harden tissues in alcohol. The sections may be stained with Löffler's alkaline methylene blue or with aniline water methyl violet or fuchsin for from one-half to one hour. After a a few seconds' contact with a 1 per cent. solution of acetic acid they are washed in water, then passed through alcohol, turpentine, or xylol or cedar oil, and mounted in xylol balsam. No stain which differentiates these bacilli very sharply from others has been found. They are decolorized when the method of Gram is employed.

Distribution of bacteria in the body.—This can be best determined by the delicate method of cultivation. In acute cases the spleen contains the largest number. In coverglass preparations of spleen pulp there may be four or five in every field of the microscope. In general, the liver contains almost as many bacteria as the spleen. The lungs, lymphatic glands, and kidneys may also contain them in moderate number. They are fewest in blood from the heart (right ventricle). In slow, chronic cases, characterized by slight ulceration of the large intestine, the number of bacteria in the internal organs is very small. From the spleen of such cases cultures are made fertile only by using bits of tissue as large as split peas.

When it is of importance to make a diagnosis from a chronic case it might be well to adopt the method suggested by Fränkel and Simmonds for typhoid fever. They wrapped the spleen in cloths wet with a solution of mercuric chloride and placed it in a warm room for twenty-four hours. The bacilli of typhoid, capable of multiplying in the organs after death when the temperature is not too low, became sufficiently numerous to be readily detected in sections. The same may be said of hog cholera bacilli. It must, however, be borne in mind that in chronic cases other bacteria may have gained entrance into the body and appear in the spleen. These, multiplying at the same time, may give rise to erroneous interpretations.

When the destruction of the mucous membrane in the large intestine is extensive, bacteria of various kinds may be found in the peritoneal cavity. The serum collected often contains several varieties of organisms, and when sections are made of the walls of the peritoneal cavity micrococci and bacilli are seen resting in a thin layer upon the peritoneum. Sometimes the pleural fluid, and still more rarely the pericardial fluid, may give rise to cultures of micrococci. These resemble the micrococci, causing suppuration in man (*staphylococci*) in their appearance and mode of growth in gelatine. The presence of anaërobic bacilli in spleen and liver has already been dwelt upon. The presence of bacteria in the peritoneal cavity is easily explained by their passage through the ulcers in the intestines. It is a fact worthy of note that only very rarely bacteria other than those causing the disease were encountered in the spleen and the circulation. Perhaps those causes or agents which destroy bacteria are less active in the serous cavities than in the blood and spleen. The various complications of chronic ulceration already mentioned, such as peritonitis, pleuritis, and pericarditis, are without doubt due to this secondary invasion of bacteria, which have the power to produce inflammation of serous membranes. It need not be said that in the earlier investigations, when little was known of bacterial diseases, the explanation of the presence of these microbes was very perplexing and misleading.

Biology of the hog cholera bacillus.—The cultivation of bacteria in nutrient media outside of the animal body serves two distinct purposes: (1) The diagnosis of specific forms so that they may be easily recognized, and (2) the study of their biology or life history.

a. Diagnostic characters.—Hog cholera bacilli are not readily distinguished from a large number of other bacteria found in surface waters and in the superficial layers of the soil, either in their form or in their manner of growth in culture tubes. The more minor differential characters we can therefore obtain, the more certain our diagnosis will be. For this end the hog cholera bacillus was cultivated in as many media as were available.

If a bit of spleen pulp from swine which have succumbed to hog cholera be thoroughly shaken up in a 10 per cent. beef infusion peptone gelatine, liquefied by a gentle heat, and the whole poured upon a sterile glass plate protected from the dust by a bell-glass and allowed to congeal, colonies of hog cholera bacilli will appear within forty-eight hours as mere specks to the naked eye. Examined under a low power they are spherical, with sharply defined border. The disk of the sphere is homogeneous without any concentric markings, and of a brownish color. This description applies to colonies beneath the surface of the gelatine layer. Colonies which grow on the surface soon spread out into thin pearly layers several times the size of the deep colonies, and roundish or irregularly polygonal in outline.

Besides the deep spherical and the flat surface colonies there is occasionally a third form present. This is a very faint cloud like colony growing between the glass plate and the gelatine, spreading as a very thin layer laterally and attaining the dimensions of a surface colony. Many such colonies have spherical prolongations upwards into the gelatine layer so as to assume the form of a lid with a knob-like handle attached to its center. It is very probable that the colony begins as a sphere in the gelatine layer near its lower surface. As soon as it touches the glass, by virtue of its enlargement, spherically, it spreads out into the attenuated, cloud-like layer. Such colonies are rare, excepting in so-called line cultures made by drawing a platinum wire across the gelatine layer before it has congealed.

Perhaps the best device for studying the growth of such colonies is the Esmarch roll culture.* The gelatine, coating the inside of the test tube, is protected from contamination and desiccation for a long time. The colonies have thus the opportunity of expanding to their utmost capacity. In a roll culture ten days old, for example, the deep colonies were about one half millimeter in diameter, perfectly spherical, the disk homogeneous, yellowish white, when viewed with a hand lens. Under a 1-inch objective it appeared reddish brown, with no markings. The surface colonies in the same tube were about eight times as large, i. e., 4 millimeters across. They presented centrally an opaque white patch or nucleus, outside of this a more translucent zone, beyond this another opaque zone, and lastly a limiting translucent border. The colonies were irregularly polygonal in outline. The alternation of thin and thick zones was very likely due to the variation in temperature to which gelatine cultures were exposed in a badly heated laboratory, for they are by no means always present. (Plate XI, fig. 2.)

Tube cultures, made by piercing the gelatine in a test tube with a platinum wire previously forced into the spleen, show minute yellowish-white spheres in the track of the wire in forty-eight hours. These rarely exceed one half millimeter even after several weeks. The surface growth spreads from the place of inoculation as a thin pearly layer of variable thickness, eventually becoming 4 to 6 millimeters in diameter; under especially favorable conditions of temperature, etc., it may become still larger. When the inoculated bacteria are very numerous the growth beneath the surface appears as a solid yellowish-white track, in which the colonies have become fused together. The gelatine is at no time liquefied. The bacilli grow somewhat larger in gelatine than in the body of diseased animals. Occasionally filaments of considerable length are met with, and in general irregular, involution forms are not

---

* These roll cultures are made as follows: The gelatine, liquefied by gentle heat in the tube in which it was sterilized, is inoculated with the tissue, blood, urine, or any liquid containing bacteria, and carefully stirred up. A second culture may be made from the first by transferring from it with a platinum loop some of the liquid gelatine. This is done when the number of bacteria in the first tube may be too numerous. A rubber cap is fitted over the mouth of the tube after replacing and trimming the cotton-wool plug, and the tube placed horizontally in ice water and rolled about its long axis until the inside of the tube is coated with a uniform layer of congealed gelatine. The tube may also be rolled on a smooth block of ice, a method which I first saw in the Pathological Laboratory of Johns Hopkins University, and which is very useful when rubber caps are wanting, or when it is desirable to keep the plug free from gelatine.

uncommon. When a bit of spleen tissue is rubbed upon an inclined surface of *agar agar* in a culture tube, isolated colonies make their appearance within twenty-four hours as circular, grayish-white, semi-translucent, very flat cones 1 to 2 millimeters across. When the bacilli are very numerous a slight prick of the spleen pulp with a platinum wire is sufficient. Otherwise the too numerous colonies will coalesce into a grayish, shining, semi-opaque layer of scarcely perceptible thickness. Its appearance on *agar* can not be distinguished from the growth of typhoid bacilli and a number of saprophytic bacteria. On blood serum the growth appears as a very thin, grayish, translucent layer. In bouillon, either with or without peptone, the bacilli grow rapidly enough at 80° to 95° F. to produce a slight opalescence within twenty-four hours. This does not increase in density subsequently. There is no membrane formed on the surface of the liquid. When standing undisturbed for one or two weeks, a white ring-like deposit of bacteria frequently forms around the tube at the surface of the culture liquid. There is only a very slight deposit formed, showing that multiplication in liquid media comes to an end within a few days.

The length of the bacilli in bouillon is about .9 micromillimeter, their width .4 to .5 micromillimeter, therefore somewhat smaller than in the spleen. When examined in a drop of bouillon suspended from the under surface of a cover-glass in a "cell," the bacteria are seen to be motile. Taken from cultures one or two days old they execute very active spontaneous movements. Their movement is one of rotation as well as translation. They quite invariably occur in pairs, and the movement of rotation is about their point of union as an axis. The pairs of bacilli as they shoot across the field have thus an oscillating motion at the same time. The bacilli do not come to rest at all, but swarm about very rapidly until the liquid is dried up.

Though this motility is most marked in recent cultures, it may still be seen at the end of one or two weeks in most liquid cultures.

The same active motion is observed in bacilli taken directly from the animal, such as the spleen of rabbits, which have succumbed to inoculation. If a bit of spleen tissue be rubbed in a drop of sterile water on a cover-glass and the whole examined as a hanging drop, in one or two minutes the bacilli become as active as in cultures.

Growth on boiled potato, when at 95° F., appears as a faint straw-colored deposit within twenty-four hours after inoclution. At 20° to 25° C., it appears one or two days later. It slowly spreads in all directions as a layer of perceptible thickness. The color changes to a dark brick red, or may remain whitish. In general the growth is darker the more rapidly the potato dries up.

The growth is also restricted by drying. In some cultures it has covered almost the entire cut surface of the potato. In others it remained as a broad band over the line of inoculation. The bacilli multiply very abundantly in *milk* without producing any alteration visible to the naked eye.

The diagnosis of hog cholera bacilli may therefore be made by taking together the following morphological and biological characters: (1) Short bacilli with rounded ends, or ovals, readily stained in aqueous solutions of aniline dyes. (2) Growth at 65° to 80° F., on gelatine, without producing liquefaction. (3) A rather feeble growth in beef infusion coupled with active spontaneous movements. (4) Growth on the cut surface of boiled potato at ordinary temperature as well as in the incubator. (5) Active multiplication in milk without any macroscopic changes. (6) Growth *in vacuo*. (7) Absence of any odor arising from the cultures. (8) Fatal effect on rabbits, guinea-pigs, and mice when inoculated. This will be discussed later on.

These characters are emphasized, since we have several times found bacteria in the internal organs of swine which may have many points in common with hog cholera bacilli; especially as regards their form, motility, and growth in gelatine. In fact in was quite impossible to decide until inoculations upon animals were made. The absence of pathogenic power was thus made the chief criterion.

(*b*) *Other physiological characters.* —Though alkaline media are as a rule most favorable for bacterial multiplication, yet there is a slight development in media containing a small amount of acid, such as Liebig's meat extract.

A feeble development was observed in sterilized hay infusion.

The temperature range of the active multiplication of hog cholera bacteria lies between 60° and 104° F., being most favorable between 85° and 100° F.

Hog cholera bacilli, though they seem to develop best in presence of oxygen, are capable of growing in what is practically a vacuum, *i. e.*, they are facultative anaërobic organisms. Comparative experiments made with such obligatory aërobic organisms as bacillus subtilis determined that while the latter shows no trace of growth in tubes from which the air has been removed, hog cholera bacilli grow quite as freely as in presence of air. When the latter are shaken up in liquid gelatine in test tubes and the gelatine rapidly congealed the colonies that appear throughout the gelatine show no difference in size, whether near the surface where air can penetrate or near the bottom of the tube.

THE DIAGNOSIS OF HOG CHOLERA BY MEANS OF INOCULATION.

The inoculation of small animals in the study of infectious diseases has been of the utmost importance from a diagnostic stand-point. Frequently bacteria which are nearly identical in form, size, and many biological characters, can only be distinguished from one another by their effect upon smaller animals. This is especially true of hog cholera, since the specific bacillus closely resembles many forms found in decomposing liquids. In the following pages, therefore, a careful exposition of hog cholera as manifested in small animals will be made to facilitate the diagnosis of this disease. During the course of these investigations rabbits have been found best for this purpose, although ordinary house mice and guinea-pigs will answer almost as well.

If a bit of spleen tissue from a case of hog cholera be inserted under the skin of the thigh, or simply rubbed upon a slight abrasion made on the inner surface of the ear, or if the bacilli from pure cultures be used, the disease will be induced. The rabbit will succumb within a period after inoculation varying from five to fourteen days (very rarely longer), depending upon the number of bacteria introduced. The period of incubation, during which the animal shows no symptoms, varies from three to five days before death. At this time the temperature rises from the normal (102.5°-103.5° F., according to the age of the animal), to 107° or 108° F., and remains at that point until shortly before death, when, if examined in time, a fall is observed. The height of the fever is accompanied by loss of appetite and a tendency to sit perfectly quiet.

The disease thus induces in the rabbit a typical continuous fever invariably ending in death. By way of illustration, the temperatures of a few cases are here reproduced :

Rabbit inoculated June 18, with $\frac{1}{8}$ cubic centimeter culture liquid; weight, $2\frac{1}{4}$ pounds ; temperature, 102.5° F. ; at end of first day, 102.7; of second day, 104.6; of third day, 105; of fourth day, 108; of fifth day, 106.4; dead on sixth day.

Rabbit inoculated June 20, with infected soil, one and one-half months old; temperature on fifth day, 103.4; on seventh day, 103.8; on eighth day, 103.8; on tenth day, 107.6; on eleventh day, 107.4; dead on twelfth day.

Rabbit inoculated June 28, with $\frac{1}{8}$ cubic centimeter of culture liquid ; weight, $2\frac{7}{16}$ pounds ; temperature, 103.7° F. ; end of first day, 103.3; of second day, 105.5; of fourth day, 107.8; found dead on seventh day.

The lesions produced are very constant and characteristic. At the point of inoculation there is a slight infiltration of the subcutis and fascia, and occasionally a slight superficial necrosis of the muscular tissue. The spleen is very large, perhaps three to five times its normal size. The liver presents on its surface yellowish white patches which correspond to one or more lobules which have undergone coagulation necrosis. These patches vary greatly in number but are rarely absent, provided the animal lived long enough to permit their formation, i. e., not less than seven to eight days after inoculation.

This necrosis appears chiefly in lobules near the surface, although occasionally the entire tissue is involved. In some livers the necrosis is complete so far as the lobule is concerned. In others it is restricted to the peripheral or portal zone of the lobule, in which case a large patch of lobules usually undergo the same change. There can be no doubt that they represent different stages in the process which seems to begin in the portal area. When sections are examined under the microscope the liver-cells show as unstained masses without nuclei. There is a remarkable absence of leucocyte infiltration around the necrosed tissue. The bacilli are very well brought out in such tissues after thorough hardening in alcohol, by staining over night in aniline water methyl violet (tubercle stain), and decolorizing slightly in $\frac{1}{2}$ to 1 per cent. of acetic acid. The deep blue bacilli appear in dense masses, chiefly in the capillaries along the edge of the necrosed area. The necrosis is most probably due to the plugging of the different vessels, thus cutting off the blood supply.

The enlargement of the spleen and the necrosis in the liver are the two important characters of the inoculation disease. In a moderate number the intestinal tract is found diseased. The submucosa of the duodenum near the pylorus is dotted with ecchymoses, which may fuse into a single hemorrhagic patch. Besides this the submucosa of the large intestine 2 to 3 inches from the rectum, i. e., the straight portion merging into the rectum, may be in the same condition. Sometimes the mucosa is beset with minute bulging hæmatomata. These intestinal lesions are due to the discharge of bacilli from the necrotic foci in the liver into the bile ducts and thence into the duodenum. The lesions in the large intestines may be influenced by the more or less prolonged stay of the feces before final discharge.

In some cases when the disease has lasted from ten to thirteen days there is a peculiar exudate in the large intestine, which may be drawn out of the anus in the form of bands or cylindrical masses 1 or 2 feet long, consisting of a translucent, elastic, whitish material, finely fibrillated when examined under a high power. The duodenum may be distended with a straw-colored semi-gelatinous mass. Hemorrhagic lesions may also be present. It seems very probable that the exudate is the result of a fibrinous or croupous inflammation of the large intestine due to the presence of hog cholera bacilli.

The lungs are occasionally the seat of hemorrhages. The kidneys contain more blood than usual, but hemorrhages are absent. Hog cholera bacilli are obtained by cultivation from nearly all the internal organs. They are usually so abundant in spleen and liver as to be detected readily in cover-glass preparations.

The following experiment shows how few bacteria are required to produce the disease :

Rabbit No.	Inoculated with culture liquid.	Remarks.
17	$\frac{1}{300000}$ c. c.	Dead on sixth day.
18	$\frac{1}{300000}$ c. c.	Dead on ninth day.
19	$\frac{1}{1000000}$ c. c.	Dead on eighteenth day.
20	$\frac{1}{4000000}$ c. c.	Dead on eighth day.

The beef-infusion culture was diluted so that one-quarter of a cubic centimeter of the liquid contained the equivalent given in the table. It s probable that No. 19 took the disease from No. 20, penned with it. The latter had ecchymoses in the duodenum, indicating that the bacilli had escaped into the intestinal canal and were being discharged with the feces. The time of death of No. 19 suggests this view.

A very small number of bacilli, therefore, are sufficient to produce in rabbits an infectious fever, or septicæmia.

The characteristic action of hog cholera bacteria on the mucous membrane is well illustrated by feeding cultures to rabbits:

September 11, 1888.—Two rabbits were starved for one day and then fed hog cholera bacilli from an *agar* culture on clover. This was readily consumed.

One rabbit was found dead on the sixth day. Agglutination of bladder to cæcum; serosa of stomach reddened. Ileum about 6 inches from valve invaginated for 2 to 3 inches. Spleen congested, slightly enlarged. Liver and kidneys engorged. Large numbers of hog cholera bacilli in spleen, as tested by the microscope and cultivation on gelatine. The other rabbit died in thirteen days. Spleen very large, congested; coagulation necrosis in liver. Lungs contained several dark-red hepatized areas. Severe lesions were found in the large intestine. The mucous membrane of lower colon and rectum were dotted with small bulging hemorrhages, and the tube contained a cylindrical mass of a material resembling colorless gelatine, partially softened in water and very elastic (fibrin?). The duodenum was distended with a pale yellowish translucent semi-gelatinous mass. Hog cholera bacilli as above.

At the same time two other rabbits were starved for a day and then 5 cubic centimeters of a liquid culture was added to 20 cubic centimeters water and given them to drink. One rabbit died in ten days. The lesions were the same as those in the rabbit just described, with this exception, that there was no coagulation necrosis in the liver. The other rabbit did not take the disease. It is quite likely that it may not have taken much of the water.

When rabbits are not easily procurable the ordinary house mice may be used. They usually appear quite well and active after inoculation until some morning they are found dead. The period of the disease is the same as that for rabbits. The spleen is very large; the liver contains many centers of necrosis. Mice have been frequently fed both with pure liquid cultures and with spleen pulp containing hog cholera bacilli. They invariably take the disease and die within seven or eight days after the first feeding, with lesions the same as those observed in mice inoculated subcutaneously. Two mice were peculiarly affected the day after eating some spleen pulp. They were scarcely able to move. Their limbs sprawled, and on being taken out of the jars they remained in this position unable to escape. They died subsequently of true infection. The symtoms recorded were due very likely to the absorption of the ptomaine produced by the bacteria in the digestive tract.

There is a remarkable regularity in the length of the disease in mice. Those inoculated at the same time and with approximately the same dose usually die not more than a half a day apart; they appear lively up to the night during which they die. In one case four mice, inoculated at the same time, all died in the night of the seventh day, though they appeared well the day previous.

The following experiment is worthy of record as indicating the effect of hog cholera virus on mice that survived infection:

October 28, 1883.—Two mice were fed with bits of liver and kidney tissue from mice which had succumbed to inoculation with hog cholera cultures. The material fed had been kept in the refrigerator for nearly four weeks in salt solution. Before feeding it was thoroughly mixed with bread crumbs and placed in the mouse cage. Both mice survived the feeding and were active December 1, over one month after feeding. One was thereupon killed with chloroform and examined. The spleen was very much enlarged, not congested abnormally. On the ventral aspect of the liver a whitish patch where diaphragm was adherent. Under the patch a large abscess. This was without doubt the site of extensive coagulation necrosis, which was now broken down.

The second mouse was killed December 19, over seven weeks after being fed. The spleen was likewise very large, the liver intact, but there were five abscesses in the left kidney, with adhesions to surrounding structures. On section one abscess was wedge shaped, the apex being at the pelvis; the whole only partly softened. These also were the result of coagulation necrosis. We have here a striking illustration of the mainly mechanical injury done by hog cholera bacilli.

The effect of hog cholera bacilli on guinea-pigs differs but little from that exerted on rabbits and mice. The duration of the disease and the lesions are the same. There seems to be a somewhat greater tolerance in guinea-pigs, although the small number used does not justify any decided statement.

Pigeons are also susceptible to this disease, but by no means to the degree witnessed in the animals just mentioned. It requires about three-fourths of a cubic centimeter of an ordinary beef infusion peptone culture to produce a fatal result (*i. e.*, about 150,000 to 3,000,000 times the dose necessary to destroy rabbits). The birds frequently died within twenty four hours after the injection, which was usually made under the skin over the pectorals on one or both sides. The pectoral after death is partly or wholly discolored, and has a parboiled appearance. The injected bacteria are present in the heart's blood and in other organs, notably the liver. If the dose is smaller than three-fourths of a cubic centimeter, the bird may survive after a week or two of pronounced illness. A large sequestrum forms in the pectoral muscle, which is gradually absorbed. Occasionally the birds will die after a week of illness manifested by diarrhea, ruffling of the feathers, and a quiet, somnolent attitude in a corner of the cage. Hog cholera bacilli are also present in the internal organs in such cases. Feeding cultures has no effect.

Fowls have been frequently inoculated and fed with cultures without showing any signs of susceptibility. Among other animals inoculated were several white rats, one gray rat, one sheep, and a calf. In none did the disease appear. In the sheep and calf a small abscess was found at the place of inoculation.

VITALITY OF HOG CHOLERA BACILLI, AND THEIR RESISTANCE TO VARIOUS DESTRUCTIVE AGENTS.

The vitality of hog cholera bacilli in cultures remains for months unchanged. The following is perhaps an extreme illustration:

A tube of beef's blood serum, coagulated by heat, was inoculated with hog cholera bacilli November 28, 1885. The platinum wire penetrated the blood serum and the bacteria grew as a slender plug in the needle-track and as a thin film on the surface. In July, 1887, more than one and one-half years later, a tube of beef infusion inoculated from the blood serum culture became promptly turbid, and contained the hog cholera bacilli only. Two mice inoculated subcutaneously with a few drops August 2, died August 8 and 9, respectively, with lesions characteristic of the disease and with the bacilli present in spleen and liver. The germs had not therefore lost their virulence. The blood serum had contracted but slightly from loss of water, there being very little evaporation from the culture tubes used.

Resistance of hog cholera bacteria to heat in liquids.—A knowledge of the degree of heat necessary to destroy hog cholera bacteria is of considerable importance, not only in its bearing upon the application of heat as a disinfectant, but also upon the various processes that are used in the preparation of pork for consumption.

Culture tubes containing about 10 cubic centimeters of beef infusion were inoculated from a culture of a certain age and then placed in a water-bath kept at the desired temperature. They were exposed for different periods of time and then removed to an incubator at 95° F. When the tubes remained clear it was inferred that the bacteria had been destroyed. A control tube was inoculated in every experiment to make sure of the vitality of the culture used.

A momentary exposure to boiling water will destroy them. When the temperature of the surrounding water is 158° F. (70° C.), the inoculated tubes remain sterile after an exposure of four to five minutes. As it takes about four minutes for the temperature of the culture liquid to reach 70° C., it is probable that a two minutes' exposure to 70° C. would be sufficient.

An exposure to 136° to 138° F. (58°–59° C.), is sufficient to destroy hog cholera bacteria in fifteen minutes. The same is true for bacteria taken directly from the spleen. An exposure to 130° F. (54.5° C.) will destroy them in one hour. Tubes exposed for one-quarter of an hour become turbid within twenty-four hours. Those exposed for one-half and three-quarters of an hour become turbid within forty-eight hours. This shows that most of the bacteria have been destroyed by such prolonged exposure. Tubes heated for one hour remain clear.

When the temperature is still more reduced, to 120° F. (49° C.), exposure for a period as long as two hours is insufficient to destroy them, although their growth may be slightly retarded.

It must be remembered that these results mean that the bacteria must be actually exposed to these temperatures for the length of time indicated. In the culture tubes employed it takes about five minutes to bring the temperature of the liquid up to 70° C., and less for lower temperatures. If, therefore, it takes longer than this for the heat to penetrate into meat or lard containing these bacteria, their destruction can not be regarded as certain within the time indicated above, and the exposure to the required temperature must be correspondingly lengthened.

Hog cholera bacteria in 10 cubic centimeters bouillon placed in a water bath at—	Time in which destroyed.
100° C. (212° F.)	Immediately.
70° C. (158° F.)	4 minutes.
58° C. (138° F.)	15 minutes.
54.5° C. (130° F.)	1 hour.
49° C. (120° F.)	Not in 2 hours.

Bacteria do not so readily succumb to heat when dried and then exposed to dry hot air. It was found that dry heat at 80° C. (176° F.) is sufficient to destroy the bacteria when exposed in a dry state for fifteen minutes.

In these experiments bacteria from cultures were rubbed on the inner side of plugged and sterilized test tubes. Gelatine was added after they had been exposed to the heat and roll cultures made. Absence of growth indicated the death of the contained bacteria.

Resistance to freezing is well illustrated in the experiments reported in the section on the vitality of hog cholera bacilli in the soil during winter.

Vitality of hog cholera bacteria in ordinary water.—The hardiness of this microbe is well illustrated by its capacity for multiplication in ordinary drinking water. To determine this, the following experiment was made:

September 8, a culture tube containing very clear Potomac drinking water* which had been sterilized several weeks previous by a temperature above 110° C., was inoculated with a platinum loop from a pure liquid culture of the bacillus. By mixing a given quantity of this water immediately after inoculation with gelatine, and making a plate culture of the same, it was found that the water contained about 26,240

* When drawn this water did not contain more than 100 to 200 bacteria to the cubic centimeter.

bacteria in 1 cubic centimeter. The water was kept in the laboratory, in which the temperature corresponded closely with that prevailing outdoors. It was examined from time to time on gelatine plates, and the number calculated for 1 cubic centimeter. The following figures give the results obtained:

September 8, 26,240 in 1 cubic centimeter (immediately after inoculation).

September 9, 201,600 in 1 cubic centimeter.

September 10, 1,296,000 in 1 cubic centimeter.

September 11, too numerous on plate to be counted.

September 13, 2,608,200 in 1 cubic centimeter.

September 15, 1,519,560 in 1 cubic centimeter.

September 17, 1,306,308 in 1 cubic centimeter.

September 29, 83,700 in 1 cubic centimeter.

October 12, 19,125 in 1 cubic centimeter.

October 21, 10,880 in 1 cubic centimeter.

November 18, 225 in 1 cubic centimeter.

December 6, a few bacteria still present, as determined by liquid cultures.

January 4, 17 in 1 cubic centimeter.

January 11, no growth on plates.

According to another experiment the vitality lasted about two months.

June 13, 1888, 10 cubic centimeters, Potomac drinking water, sterilized in a Salmon culture tube, was inoculated with a platinum wire to which a minute bit from the surface growth of an *agar-agar* culture adhered. Plate cultures prepared immediately after thoroughly shaking the tube, indicated that each cubic centimeter of the water contained from 1,000,000 to 2,000,000 germs.

Plate culture of June 14 shows in $\frac{1}{30}$ cubic centimeter of water a very large number of colonies. Roll cultures made June 22, July 3, and July 16 contain a smaller number.

A roll culture of August 4 contains about 200 colonies (*i. e.*, about 6,000 in 1 cubic centimeter).

A roll culture of August 25 remains sterile. A tube of beef infusion to which about 1 cubic centimeter of the water had been added September 15 contained a large coccus five days later; no hog cholera germs.

These cultures show that the bacilli perished in about two months. The difference in the results obtained in this and the last experiment may have been due to the season of the year.

That the bacilli can be kept alive in clear river water for from two to four months and perhaps longer is a fact very significant in itself. When we consider, moreover, that the added bacteria in the first experiment multiplied so that each individual was represented by ten at the end of five days, the hardiness of the bacillus is very evident. The danger from infected streams into which feces from sick animals find their way is thus proved beyond a doubt. Stagnant streams and pools are more dangerous, since the water is but slowly renewed, while in rapidly flowing streams the bacteria are speedily carried away. On the other hand, the latter may spread an epidemic from one place to another.

Resistance to continuous drying at ordinary temperature (60° *to* 80° *F.*— A number of experiments were made to determine this point. Some of them are reported in full in preceding Annual Reports of the Bureau, and are therefore simply summarized in this place.

(1) A series of cover-glasses, upon each of which a drop from a liquid culture had been dried, were placed in bouillon at different times. No growth in bouillon after they had been dried from seven to nine days.

(2) Minute bits of spleen tissue from a pig which had succumbed to hog cholera were dried on cover-glasses. These were capable of infecting bouillon up to the twenty-third day.

(3) Spleen tissue dried on four cover-glasses gave rise to pure cultures of hog cholera bacilli after forty-nine days.

(4) Hog cholera bacteria from a liquid culture one week old, dried on silk threads, were placed on a gelatine plate from time to time to observe any growth. They were still capable of development after twenty-one days, when the stock of threads was exhausted.

(5) Threads steeped in a liquid culture one day old and dried were placed on gelatine plates, as described in series 4. No colonies appeared on the twenty-seventh and twenty-eighth days. A few appeared later on, showing that even after thirty days a few still survived.

In these series of experiments the vitality of the bacilli was not exhausted after forty-nine days in one series; in another it was destroyed in less than ten days.

The following series of experiments, made during the present year, show how very varied is the length of time during which the bacilli remain alive when dried:

June 15.—Minute bits of spleen tissue, from a rabbit which had died of hog cholera after inoculation with a pure culture, were rubbed on sterile cover-glasses and kept under a flamed funnel, plugged with cotton-wool.

On July 3, 6, and 9 two cultures were made by dropping into each tube of beef infusion a cover-glass. All six tubes remained clear. The bacilli had thus perished within three weeks.

In the following experiment hog cholera bacilli remained alive for more than four months:

June 8.—From an *agar-agar* culture, three days old, some of the abundant surface growth was placed upon sterile cover-glasses, each receiving about as much as the point of a platinum wire could hold. These were placed on a flamed glass support under a flamed and plugged glass funnel, and kept in the laboratory, the air of which was moderately dry.

Two cover-glasses dropped into beef infusion peptone June 25. One culture remains clear, the other clouded in one day; contained only motile hog cholera bacilli.

Two covers were added to beef infusion June 29, July 9, and July 16. All six cultures became clouded and contained only hog cholera bacilli.

Of two cultures made July 23 only one becomes clouded, and is a pure culture of hog cholera bacilli. Of two cultures made August 2, both become clouded with hog cholera bacilli only.

Four cultures, made on August 9, 22, 30, and September 14, became clouded with the same microbes, and these alone. A tube into which a cover-glass had been dropped September 21 remained sterile. A tube inoculated in the same way September 26, became turbid on the second day, and contained hog cholera bacilli only. A tube inoculated on October 8 was still fertile. The stock of cover-glasses had become exhausted.

It will be seen from the above results that some of the germs were dead in one and a half and three and a half months; all the rest were capable of multiplication for four months. The prolonged vitality was no doubt due to the massing together of the germs from the *agar* culture, for on those cover-glasses which failed to inoculate cultures, or in which the appearance of growth was retarded, the quantity of growth was smallest. The interesting question here arises whether it is the oxygen of the air which gradually destroys the bacilli, since this is more or less kept away when they are massed together, and since all previous experiments with bacilli suspended in liquids have shown that the period of their vitality may average not more than two months when in a dry condition.

On August 30, after having been dried for two months and twenty-two days, two mice were inoculated with a liquid culture obtained from one of the cover-glasses. Both died of hog cholera on September 14 and 15, respectively.

The vitality of hog cholera bacilli during continuous desiccation may thus last from two weeks to more than four months.

In the soil, and in nature generally, bacteria are rarely subjected to continuous drying, but to alternate wetting and drying. In order to observe the effect of such alternation, some of the same *agar-agar* culture used in the preceding experiment was placed in the bottom of a sterile watch-glass under a funnel June 8.

June 15.—About one-third cubic centimeter of sterilized distilled water added to the watch-glass, so as to cover the dried film completely. The water was evaporated entirely next day.

June 22.—Sterile water added again; dried up next day.

July 3.—Water added again and two cultures made; both remain sterile.

July 6.—A liquid culture made by adding some sterile water to the dried culture mass, stirring it up and transferring the water with a sterile pipette to a tube of beef infusion. This also remained clear.

Thus bacilli from the same culture which resisted continuous drying for at least four months were destroyed in less than a month when a moist and a dry state alternated. This fact, so striking and important, needed confirmation.

September 15.—A considerable quantity of the surface growth from an *agar-agar* culture two days old was rubbed on the bottom of a sterile watch-glass covered by a plugged funnel as before. Thoroughly dry next day.

September 18.—A small quantity of sterile distilled water added. This was dried up next day; the germs had spread out into a thin layer.

September 21.—Sterile water added again and a tube of beef infusion peptone inoculated with a loop from the liquid stirred up. Tube turbid with hog cholera bacilli next day.

September 26.—Sterile water added again. The tube of beef infusion peptone inoculated at the same time. Pure culture next day.

October 1.—The same process repeated; culture contains hog cholera bacteria.

October 8.—Dried bacilli wetted again with sterile water. The culture made from them becomes turbid; only hog cholera bacilli present.

October 15.—The same process repeated; the inoculated tube becomes turbid after forty-eight hours.

October 22.—After wetting the dried growth again, it was thoroughly stirred up and several drops added to a liquid culture. This tube remained permanently clear.

This experiment therefore confirms the other in showing that hog cholera virus is far more quickly destroyed when it is alternately moistened and dried than when it remains continuously dry. In both tests the germs died in one-third the time required to destroy dried virus.

SOME EXPERIMENTS ON THE LENGTH OF TIME DURING WHICH HOG CHOLERA VIRUS REMAINS ALIVE IN THE SOIL.

The virus of hog cholera is quite tenacious of life in spite of the fact that no spores are formed. During the past year some preliminary experiments were made concerning the vitality of hog cholera bacteria in the soil. This becomes infected during epizootics of this disease by the discharges of the sick perhaps more thoroughly than anything else in the surroundings of the animals. Moreover, it is the most difficult to disinfect, as we have no knowledge of the depths to which the living virus may be carried by water. If it can be shown that the life of such virus in the soil is speedily destroyed, the precautions to be taken would be quite different from those needed if it exists for a long period of time.

Experiment 1.—A small flower-pot containing soil was sterilized by moist heat at 110° to 118° C., and protected from drying and dust by a large bell jar. On its surface about 100 cubic centimeters of a bouillon peptone culture of hog cholera bacteria was poured and the whole maintained moist and at the laboratory temperature. The soil used was a very fine loam from the grounds of the Department of Agriculture. The bulk of the soil consisted of grains not larger than $\frac{1}{100}$ millimeter ($\frac{1}{2500}$ inch). Roll cultures from the soil after a few days demonstrated the presence of immense numbers of bacteria. From this soil rabbits were inoculated from time to time by stirring up a little soil in some sterile beef infusion and injecting the clear supernatant liquid hypodermically. The soil was infected September 17, 1887. The appended table gives the inoculations into rabbits to test the virulence of the soil. The rabbits which succumbed died of hog cholera, as indicated by the lesions and the bacteriological examination.

No.	Date of inocu- lation.	Time after in- fection of soil.		Died.	Remarks.
		Mos.	*Days.*		
1	Oct. 10		25	Oct. 17	Enlargement of spleen; coagulation-necrosis in liver; hemorrhagic lesions in duodenum and rectum, in lungs and heart; numerous hog cholera bacteria in spleen and liver.
2	Oct. 18		31	Oct. 24	Spleen and liver and lungs as in No. 1.
3	Nov. 4		46	Nov. 12	Spleen, liver, duodenum, and rectum as in No. 1.
4	Dec. 12	2	25	Dec. 23	Spleen large. pale; coagulation-necrosis in liver slight; no other lesions; hog cholera bacteria as in No. 1.
5	Jan. 9	3	23	Remains well.....	
6	Jan. 23	3	23	...do	

The infectious quality of this soil when a month old was tested on two pigs, by feeding each directly with a tablespoonful of the soil. One showed no disease; the second, a young pig, became very sick and was killed on the eighteenth day, being unable to rise.

Autopsy.—Animal about eight weeks old, very thin. No skin lesions. Lungs normal, excepting the cephalic and ventral lobes of left lung, the ventral and the root of principal lobe of right lung. These are airless, bright red, mottled with yellowish points. indicating broncho-pneumonia. The spleen was small. The lymphatic glands of large intestine very large, tough, whitish. The walls of cæcum and colon over one-eighth inch thick; do not collapse when slit open. The mucosa is everywhere covered by a firmly adherent yellowish-white slough, extending as far as rectum. The ileum for about 2 feet from lower end has the mucosa likewise involved in superficial necrosis, but only on the summit of the longitudinal folds. In the stomach a portion of the fundus is covered by a friable deposit made up chiefly of large granular cells with single nucleus. Gelatine and liquid cultures from heart's blood and spleen contain only hog cholera bacteria. The gelatine cultures indicate only a moderate number of colonies. The bacteria had also penetrated into the diseased lung tissue. They were obtained on plate cultures. and a rabbit inoculated subcutaneously with some beef infusion in which a bit of lung tissue had been torn up died in eleven days with coagulation necrosis in liver, enlarged spleen. and numerous hog cholera bacteria in both organs. The broncho-pneumonia may have been due to the aspiration of some of the soil during the forced feeding.

It has thus been shown that moist soil, not allowed to dry out and kept in a summer temperature ranging from 60° to 95° F., retained its virulence for rabbits from two to three months. At the end of this time fungi and other bacteria had found their way into the pot of soil, as shown in roll cultures. This and other reasons drawn from observations of this germ lead to the conclusion that its life becomes extinguished with its pathogenic effect on rabbits. This phase of the question is not to be overlooked, for even if a germ should no longer prove pathogenic, but remain alive. it is not improbable that it may regain its original virulence under certain unknown circumstances.

15612 H C——6

The severe disease caused in the young pig with soil which had been infected one mouth ago shows the great susceptibility of young animals to this disease. Since rabbits are most susceptible of any animals to hog cholera virus, it was not thought necessary to experiment on pigs after the former failed to take the disease.

It may be argued that the conditions of this experiment were abnormal, in that the bacteria were not exposed to the various meteorological conditions, such as rain and drought, freezing and thawing, and the competition with other bacteria in the soil. This is true. The conditions just enumerated are opposed to such a long life in the soil. A condition favoring it, on the other hand, is the low temperature in winter, which may act as a means of preserving the life of bacteria.

These objections were partially removed by the following experiments:

Experiment 2.—A small pot, about 5 inches across the top, filled with soil, is placed in another pot of twice the diameter, also filled with soil in such way that the rims of both are on the same level. In this position they were sterilized for three and one-half hours, at a temperature of about 110° C., in a steam sterilizer. The outer pot was to protect the inner one to some extent from coming in direct contact with the garden soil in which it was subsequently placed. About 200 cubic centimeters of a beef infusion culture of hog cholera bacteria three days old was ultimately mixed with the soil of the inner pot by removing the upper layers and pouring the culture liquid upon the deep layers and then upon the surface, after replacing the superficial layers. The pots were then placed in the grounds of the Agricultural Department so that the top was level with the surrounding soil. Nothing was placed over the surface to protect them from contamination with ordinary bacteria.

The infected soil was placed in the grounds December 16, 1887 During the remainder of December the out-door temperature oscillated slightly above and below the freezing point, remaining for several days below this point at the end of the month. The soil in the pot was saturated with water during part of the winter and the surface covered with it. It had probably become tightly packed, and hence impervious after the culture liquid had been poured upon it.

The appended table indicates the persistence in virulence of the infected soil as tested upon rabbits. The soil was taken from the surface of the pot, or from near the bottom as indicated, and the inoculations made as in the preceding experiments:

83

No.	Date of inocula- tion.	Soil taken from—	Number of days after infection of soil.	Rabbit dies—	Remarks.
1	Jan. 5	Surface	20 days	January 16 ...	One lobe of liver involved in coag- ulation-necrosis bacilli of hog cholera present.
2	Feb. 1	...do	1 month 13 days ..	February 18 ..	Spleen very large, dark; coagula- tion-necrosis in liver; numerous hog cholera bacilli in both organs.
3	Feb. 7	Depth	1 month 19 days ..	....	Killed February 27; not diseased.
4	Feb. 23	Surface	2 months 7 days ..	March 5	Spleen and liver as in No. 2; hem- orrhagic lesions in duodenum and rectum; hog cholera bacilli present.
5	Mar. 5	.. do	2 months 18 days	Remains well	
6	Mar. 21	.. do	3 months 5 days ..	....do	

Rabbit No. 3 was inoculated with soil from beneath the surface. This was obtained by removing the inner pot, thrusting a small metallic cork-borer previously flamed through the hole in the bottom of the pot upwards, about half-way to the surface. The soil brought down in the inside of the borer was used for inoculation. In this way it was ob- tained unmixed with soil from the superficial layer. The rabbit re mained well for twenty days after inoculation. This is longer than any period of incubation which we have thus far observed. Owing to ex- tensive disease of the external ear caused by psoroptes (lice), it was killed on the twentieth day. There were no indications of hog cholera. The deeper layers of the soil had thus rid themselves of the infectious material sooner than the surface. Roll cultures showed the presence of a large number of pearly colonies spreading rapidly and made up of non-motile bacilli.

Rabbit No. 2 lived seventeen days after inoculation. Was this long period of incubation due to an attenuation of the virus, or to the scar- city of the surviving bacteria? The lesions were practically identical with those usually found. In order to learn whether attenuation had actually taken place, a rabbit was inoculated subcutaneously with three to four drops of a beef infusion culture prepared from the blood of No. 2. It died in five days with extensive coagulation-necrosis in liver; spleen very much tumefied and petecchiæ in rectum. There was no at- tenuation, therefore, so far as the second rabbit is concerned. There still remains the possibility that the bacteria, attenuated when inocu- lated into the first rabbit, grew in virulence during the long period of incubation until they had attained their original pathogenic power. when the rabbit died.

Experiment 3.—February 20, two pots of soil, one within the other, were sterilized as in the preceding experiments. The soil of the inner pot was saturated from below up with 100 cubic centimeters of a beef infusion culture one day old, of hog cholera bacteria. The whole was buried in the garden on a level with the soil and not protected in any way. Rabbits were inoculated with the soil as previously indicated.

[Soil infected February 20.]

No.	Date of inocula-tion.	Soil taken from—	Number of days after inoculation.	Remarks.
1	Mar. 10	Surface	18 days	Rabbit dies on eleventh day. Spleen very large; coagulation necrosis in liver; hog cholera bacilli present.
2	Mar. 29	Depth	38 days	Rabbit died on eighth day ; lesions as in No. 1.
3	.. do	Surface.....	...do	Rabbit remains alive.

Soon after the last inoculation the pot of soil was accidentally removed by laborers working in the garden, so that the experiment came to a premature close. It shows, however, that the bacteria in the depths of the soil were alive thirty-eight days after inoculation, while at this same time they were dead in the surface layers.

Experiment 4.—Soil prepared, sterilized, and infected precisely as in the preceding experiment, was buried in the garden April 4, 1888. The soil soon became packed hard and dry on the surface to a depth of one-half to 1 inch. The following table gives the inoculations:

[Soil infected April 4.]

No.	Date of inocula-tion.	After soil infection.	Soil taken from—	Remarks.
1	Apr. 25	21 days	Surface	Dies April 30; enlarged spleen. Slight necrosis in liver.
2	...do ...	...do	Depth	Dies May 2. Same lesions as in No. 1.
3	May 12	1 month 8 days ...	Surface	Remains alive.
4	...do ...	...do	Depth	Dies May 24. Same as No. 1, besides hemorrhagic lesions in rectum.
5	May 29	1 month 25 days ..	Surface	Remains alive.
6	...do	...do	Depth	Dies June 6. Same lesions as in No. 1.
7	June 15	2 months 11 days..	....do	Dies June 29; spleen and liver lesions as in No. 4.
8	...do	...do	... do	Remains well.
9	July 11	3 months 10 days..	... do	Do.
10	...do	.. do	....do	Do.

The presence of hog cholera bacteria was demonstrated by cultivation from spleen and heart's blood in all the animals that died.

Experiment 5.—Soil sterilized and infected with hog cholera bacteria as in preceding experiments. Placed in the department grounds May 18. June 20, surface layer dry and very hard ; boring necessary to get some of it out. Two rabbits inoculated from surface and deep soil. The result of these and subsequent inoculations is given in the following table :

[Soil infected May 18.]

No.	Date of inoculation.	Length of time after soil infection.	Soil taken from —	Remarks.
1	June 20	33 days	Surface.....	Dead July 4 (hog cholera).
2	...do ....do		Depth	Remains well.
3	July 20	2 months 2 days...	Surface....	Do.
4	.. do ....do		Depth	Do.

Experiment 6.—July 14 a pot of soil was sterilized and infected throughout with 100 cubic centimeters of a culture of hog cholera bacilli as before and placed in the Department grounds.

No.	Date of inoculation.	Length of time after soil infection.	Soil taken from--	Remarks.
1	Aug. 10	27 days	Surface.....	Dies August 23; spleen very large. Necrosis in liver; hog cholera bacilli in spleen; intestines diseased.
2	Aug. 10	... do	Depth	Died August 14; spleen and liver as in No. 1.
3	Aug. 24	1 month 10 days ..	Surface.....	Remains well.
4	Aug. 24	... do	Depth	Do.

These various experiments extended over the coldest as well as the hottest period of the year, and may therefore lay claim to more or less completeness. They may all be brought together in the following table:

No.	Date of soil infection.	Rabbits failed to take the disease when inoculated—	
		With surface soil after—	With deeper soil after—
1	Sept. 17	3 months 23 days	
2	Dec. 16	2 months 7 days	1 month 19 days.
3	Feb. 20	1 month 7 days	Not completed.
4	Apr. 4	1 month 25 days	3 months 10 days.
5	May 18	2 months 2 days	33 days.
6	July 14	1 month 10 days	1 month 10 days.

It may be said in general that hog-cholera germs will perish in the soil in from two to four months, depending on the season, moisture, and depth from the surface. In what direction these three factors influence their vitality the experiments are not complete enough to show. We may, however, safely assume that frost has no rapidly destructive effect upon them, while drying (experiments 5 and 6) seems much more destructive. Attention has already been called to the destructive effect of alternate wetting and drying. So far as the above results go, a period of at least four months should be allowed for the natural destruction of hog cholera virus in the superficial layer of the soil, *i. e.*, in a layer extending about 2 or 3 inches below the surface.

A FEW ADDITIONAL OBSERVATIONS AS TO THE VITALITY OF HOG CHOLERA VIRUS IN THE SOIL AT THE EXPERIMENTAL STATION.

(1) A pen has a concrete floor, with a gentle slope towards the back. From January to December, 1886, at least 70 pigs died in this pen from hog cholera. During this time all the liquid in the pen drained into the field back of it. During February and March of 1887 a number of pigs in this pen died of swine plague. Two survivors from the disease were removed September 15, leaving the pen empty. In December, 1887, the infected ground behind the pen was inclosed, making a yard 6 feet deep and 16 feet long and communicating with the pen. December 15, 1887, 3 pigs about two months old were placed in the concrete pen and allowed to run in the yard behind it. All three continued well. To make sure that there were no ulcerations escaping observation one of the pigs was killed one and a half months after their transfer to this pen. There were no lesions indicative of hog cholera; cultures from the spleen remained sterile. The two remaining pigs were well three and a half months later.

These observations show that ground thoroughly infected for more than a year was free from infectious properties eight months after the death of the last case.

(2) Upon a plot of ground of one-fourth to one-third acre about 100 pigs which had died of cholera during 1886 were buried, being covered with 1½ to 2 feet of soil. During 1887 no animals were buried here, but many had been interred within 3 to 8 rods of this plot during the latter half of the present year. A triangular yard 16 feet on each side was fenced off upon the old ground, with a movable pen in one corner for shelter.

December 15, 1887, 3 pigs were penned in this yard. They remained well; one and a half months later 1 was killed and found healthy. The two remaining ones were well after three and a half months.

This experiment shows that ground containing the bodies of numerous animals which had died of hog cholera was not infectious after one year.

(3) The following notes are valuable in illustrating how rapidly the ground may be freed of infectious matter:

A yard 6 by 10 feet was enclosed by a fence and made to communicate with a double pen having concrete floors. This pen had been infected by pigs from the outbreak described on page 37. The soil of the yard is a clay loam.

November 27.—Nos. 434, 435 now occupy the pen and Nos. 436, 437 added to-day.

November 28.—Nos. 434, 435 both die of hog cholera. These had been infected by feeding (see p. 50 for notes on these and other animals in this pen).

December 10.—Viscera from No. 42 fed to Nos. 436, 437.

December 12.—Viscera from No. 43 scattered over the soil of the yard, readily eaten by Nos. 436, 437; later in the day Nos. 439, 440, and 443 added.

December 17.—Viscera from No. 47 and No. 48 scattered over the ground, and Nos. 448 to 453 inclusive placed in the yard.

December 27.—No. 443 dies of cholera.

December 28.—No. 448 dies of acute cholera.

December 29.—Nos. 436, 439, 440, 449, and 452 all die of acute cholera. The pen may now be said to be thoroughly infected.

January 1, 1888.—No. 437 found dead.

January 3.—Nos. 450, 451, and 453 removed.

January 10.—After the yard and pen have been vacant for a week and simply brushed out but not disinfected nor cleaned thoroughly two healthy pigs (Nos. 479, 480) are placed in it to determine whether the ground is still capable of giving the disease.

During the greater part of January the ground was frozen over, but during February and March there are frequent, prolonged thaws.

March 29.—Nos. 479, 480 have remained well since January 10. They are now removed. The yard itself is now converted into a genuine mud-bole.

March 30.—Two fresh pigs are transferred to this yard.

July 24.—They have remained well for nearly four months.

This virus was in the first place either destroyed or made harmless by the prolonged frosts in January, so that the pen and yard were thoroughly disinfected by natural agencies and perfectly safe two months later. The disease did not re-appear in the following midsummer. It may also be seen from the notes that the two pigs placed in the infected pen but one week after the removal of diseased and infected animals did not take the disease, perhaps because the infectious matter was frozen up and slowly killed in this condition before they could get at it.

THE EFFECT OF SOME DISINFECTANTS ON THE VIRUS OF HOG CHOLERA.

In the report for 1886 there are given *in extenso* a series of experiments to test the destructive power of the more reliable disinfectants on hog cholera bacilli from liquid cultures usually from one to two days old. The results, owing to their practical importance, are also summarized here.

The method employed needs a few comments. A few drops of culture liquid were added to the disinfectant solution in a sterilized watch glass under a bell glass. After certain regular intervals of time platinum loops holding about $\frac{1}{100}$ cubic centimeter were used to transfer this liquid to tubes of beef infusion. These were placed into a thermostat and watched for a number of days. The loop of disinfectant carried over into 10 cubic centimeters of sterile beef infusion is diluted to such an extent (about 1,000 times) as to be practically of no account whatever. This method is more sensitive than the method originally employed by Koch. He used bits of silk thread impregnated with the specific organism to be tested and placed them upon plates of gelatine to develop.

It has been suggested that hog cholera bacteria may survive in the brine from salted meats long enough to be exported to other countries and communicate the disease. It need not be said that this is impos sible, for the attenuating action of a concentrated salt solution would be in itself sufficient to speedily destroy the pathogenic power of the bacilli, even if their life were not destroyed. The following experiment shows, however, that they are absolutely killed within four weeks:

Ten cubic centimeters of a saturated solution of common salt was added to each of three culture tubes. These were then sterilized and 1 cubic centimeter of a fresh bouillon culture of hog cholera bacilli mixed with the contents of each tube. Small quantities of liquid from one of these tubes were introduced into sterile bouillon every two or three days. Up to the twenty-first day the bouillon became clouded, indicating that the bacilli were still alive in the brine. After the twenty-ninth day the bouillon remained clear. The two remaining tubes of brine hitherto untouched were also found sterile when exam ined by means of cultivation a day or two later.

Mercuric chloride was found destructive to the bacillus when diluted in the proportion of $1 : 75000$ (.001¼ per cent.).

Several drops of a culture were mixed with about 1 cubic centimeter of a .1 per cent. solution, and tubes inoculated from this at the end of two, four, six, eight, and ten minutes. Tubes remain sterile. To show that the antiseptic effect of the liquid transferred with the platinum loop was *nil*, one of these tubes was inoculated again from another cul ture. This tube was turbid on the following day.

Five tubes treated in the same way with .05 per cent. solution. All remain sterile.

Five tubes inoculated with bacteria exposed for the same periods of time to a .01 per cent. solution. All remain clear.

Five tubes treated as before, using a .005 per cent. solution. Perma nently clear.

Five tubes treated as before, using a .002 per cent. solution. All tubes clear, excepting the one inoculated after six minutes' exposure.

Five tubes inoculated at the end of five, ten, fifteen, twenty, twenty five, and thirty minutes after exposure to a .001 per cent. ($1 : 100000$). Tubes inoculated after five and ten minutes turbid next day. On the second day all but the one inoculated after thirty minutes turbid and containing pure cultures of the bacterium.

The limit of disinfection for this period of time must therefore lie between $1 : 500000$ and $1 : 100000$; hence five tubes were inoculated as above, using a solution of $1 : 75000$ at the end of seven, ten, fifteen, twenty, twenty-five, and thirty minutes. All tubes remained clear.

Carbolic acid destroys the bacillus in solutions containing from 1 to 1¼ per cent. of the acid by weight.

Five tubes inoculated after treating bacilli from a liquid culture with a 1 per cent. solution for five, ten, fifteen, twenty, and twenty-five min utes. All turbid on the following day. The two last tubes were also examined on gelatine plates and the cultures found pure.

With a 2 per cent. solution five tubes inoculated after ten, fifteen, twenty, twenty-five, and thirty minutes remained sterile. The same result with a 1½ per cent. solution. With a 1¼ per cent., tubes inocu lated at the end of seven, ten, fifteen, twenty, twenty-five, and thirty minutes remained clear, excepting the first, which contained *bacillus subtilis.*

Passing to a one-half per cent. solution, tubes inoculated at the same intervals became turbid with the bacterium sown. With a three-fourths per cent. solution the result was the same.

Passing back to a 1 per cent. solution, tubes inoculated at the same intervals remained sterile.

There seems to be an incompatibility between the first and last series. If we examine the others, however, we must conclude that the limit of disinfection lies between 1 and 1¼ per cent.

Iodine water was prepared by shaking up some iodine in distilled water, which assumed an amber tint. This solution destroyed the ba. cillus in fifteen minutes, as the following experiment shows:

Six tubes were inoculated with bacilli after they had been exposed to the action of the iodine water for seven, ten, fifteen, twenty, twenty-six, and thirty-one minutes. On the following day the first tube became turbid; on the second the ten-minute tube was turbid and found to be a pure culture of the bacilli sown. The other tubes remained sterile. One of the tubes, inoculated later, showed its capacity for sustaining growth by becoming promptly turbid.

Permanganate of potash. —A series of experiments with this substance, conducted in the manner described above, showed that the bacillus is killed by fifteen minutes' exposure to .02 per cent. solution (1 : 5000).

In order to obtain this result a 5 per cent. solution was tried first. Tubes inoculated after an exposure of the virus for seven, ten, fifteen, twenty, twenty-five, and thirty-one minutes remained permanently clear. One of these tubes, subsequently inoculated with the unaffected virus, was turbid next day. Two and a half per cent., 1 per cent., one-half per cent., one-fourth per cent., one-tenth per cent., and one-twentieth per cent. solutions were tried in the same way. The six tubes used for each solution remained sterile. Finally, a one-fiftieth per cent. (1:5000) was used. Tubes were inoculated after an exposure of the virus for two, four, six, ten, fifteen, twenty, twenty-five, and thirty minutes. On the following day the four first tubes were turbid; the fifth and seventh remained sterile; the sixth and eighth contained a fine bacillus. These two tubes, as was found later, belonged to a lot which, through some carelessness, had not been properly sterilized, and the majority became turbid before use.

Mercuric iodide was found to destroy the bacillus in solution of 1:1000000 in ten minutes.

Two grams of potassium iodide and 1 gram of mercuric iodide were dissolved in 100 cubic centimeters of distilled water, making a 1 per cent. solution of the disinfectant in a 2 per cent solution of potassium iodide. This solution, diluted with sterile distilled water so as to make .1 per cent., killed the bacilius of hog cholera taken from liquid cultures in less than five minutes; .01 per cent. (1 : 10000), .002 per cent. (2 : 100000), .001 per cent. (1 : 100000), and .0005 per cent. (5 : 1000000) destroyed the germ within two minutes.

When the solution was diluted so as to make .0002 per cent. (2 : 1000000) and .0001 per cent. (1 : 1000000), it was found that with both solutions tubes inoculated with bacilli, after an exposure of two and five minutes, were opalescent, the bacilli introduced having multiplied, while the remaining tubes (ten to thirty minutes) were sterile. These two solutions, therefore, were still powerful enough to kill the germ in ten minutes. The dilution had been carried so far as to make them practically equivalent in disinfecting power.

Sulphate of copper.—This disinfectant, which seems to be more effective than most other metallic salts, was tried in solutions containing 2 per cent., one-half per cent., one-tenth per cent. Both the 2 per cent. and the one half per cent. solutions destroyed the germ within five minutes. Tubes inoculated with bacilli after an exposure to the one-tenth per cent. solution for five, ten, and fifteen minutes, became turbid; those inoculated after an exposure of twenty, twenty-five, and thirty minutes remained clear.

The disinfectant power for short periods of time may be said to lie between one-half and one-tenth per cent. In this, as in other tests, one or two drops of the culture were added to 5 cubic centimeters of the disinfectant. A slight flocculent precipitate formed each time.

Of *hydrochloric acid* a .2 per cent. solution of the acid, made by adding 4.2 cubic centimeters of chemically pure acid (containing about 40 per cent. HCl) to 95.8 cubic centimeters of water, destroyed the germ in less than five minutes.

Chloride of zinc.—A 10 per cent. solution of this salt failed to destroy the vitality of the bacilli in ten minutes; 20 cubic centimeters of (Squibbs) chloride of zinc, containing 50 per cent. of the salt, were added to 80 cubic centimeters of sterile distilled water, to make a 10 per cent. solution. A drop from a culture five days old was mixed with 6 cubic centimeters of this solution, from which mixture tubes were inoculated at the end of five, ten, fifteen, twenty-five, and thirty minutes. The two first tubes became clouded.

Sulphuric acid.—A .05 per cent. solution (1 : 2000) was fatal to the bacilli of hog cholera in less than ten minutes.

Without going into detail, it is sufficient to say that the results were reached as indicated above. Tubes containing sterile beef broth were inoculated at the end of five, ten, fifteen, twenty, twenty-five, and thirty minutes with bacteria exposed to one-half per cent. and one fourth per cent. No development. Those inoculated with one-fortieth per cent. became clouded, each being a pure culture of the bacillus inoculated. When one-twentieth per cent. was tried only the five-minute tube became clouded. The solution (by weight) was made from sulphuric acid containing 96 per cent. of the acid (specific gravity 1.838).

It must be remembered that the foregoing tests were made upon bacteria in an active, vegetative state. It is probable that in the dried condition it would have taken solutions of the same strength somewhat longer time to destroy their vitality. To briefly summarize the results, placing the least effective substance first, we obtain the following table :

Chloride of zinc in a 10 per cent. solution destroyed the bacilli in liquid cultures in fifteen minutes.

Carbolic acid, 1 to 1¼ per cent. (1:100), in five minutes.

Iodine water, in fifteen minutes.

Hydrochloric acid, one-fifth per cent. (1:500), in less than five minutes. (Only a .2 per cent. solution of this acid tried.)

Sulphate of copper, one-tenth per cent. (1:1000), in fifteen to twenty minutes.

Sulphuric acid, one-twentieth per cent. (1:2000), in less than ten minutes.

Permanganate of potash, one-fiftieth per cent. (1:5000), in fifteen minutes.

Mercuric chloride, one seven hundred and fiftieth per cent. (1:75000), less than five minutes.

Mercuric iodide in one ten-thousandth per cent. (1:1000000), in ten minutes.

The above table would without doubt be very materially altered if the test had been made with virus mixed with considerable organic matter. This is very well illustrated by the experiments on lime, crude carbolic acid, and additional experiments on sulphuric acid, given in full in the following pages. Mercuric chloride, for instance, often fails completely in albuminous liquids, although it is one of the best destructive agents of bacteria suspended in water. So permanganate of potash, which would have little or no disinfectant power in liquids highly charged with organic matter, has a very high power in this direction in the experiments detailed above.

The only substances in the above list, which, in our estimation, would be useful for purposes of disinfection, are mercuric chloride and carbolic acid. Of the former, more will be said in discussing the subject of prevention. Carbolic acid as it usually appears in the market is too expensive to be used on a large scale.

CRUDE CARBOLIC ACID.

Laplace (*Deutshe Med. Wochenschrift*, 1888, 121), found that the so-called crude carbolic acid, which will not dissolve in water, is capable of solution when mixed with an equal volume of commercial sulphuric acid. A 4 per cent. solution in water of this mixture was sufficient to kill anthrax spores within forty-eight hours, and a 2 per cent. solution destroyed them within seventy-two hours. Pure carbolic acid in a 2 per cent. solution has no effect on anthrax spores. The crude material contains about 25 per cent. of carbolic acid. Experiments were made to test the germicide effect on hog cholera bacilli. The crude carbolic acid used was a very dark, reddish, thick liquid, smelling strongly of tar, and not translucent even in a layer one-half inch deep. When a drop of this liquid was added to water it broke up into a few globules, which settled to the bottom without undergoing solution. When an equal volume of commercial sulphuric acid was added and the mixture thoroughly shaken, a few drops added to water caused a faint turbidity, but none remained in a globular condition. When more was added the water assumed a grayish, opalescent appearance, similar to water in which a small quantity of soap is dissolved.

To test this mixture, a culture liquid was prepared like that used in testing the disinfecting power of slaked and unslaked lime. (See p. 96.) Beef infusion (to which the white of an egg was added in the proportion of one egg to 600 cubic centimeters of the infusion) was neutralized and sterilized without previous filtration at 110° C. in Erlenmeyer flasks, containing each about 150 cubic centimeters. This turbid liquid, with its large quantity of flocculent material, was inoculated with hog cholera bacilli and allowed to stand for a week. Each flask then received a certain quantity of the mixture and was thoroughly shaken up. After certain intervals of time a platinum loop of liquid was withdrawn from each flask, mixed with gelatine, and a roll culture made to indicate the number of surviving bacilli. To make sure of the vitality of the bacilli in the culture a control culture was made from one of the flasks before adding the disinfectant.

Flasks with 150 cubic centimeters culture-liquid, to which was added of the mixture of carbolic and sulphuric acids—	Number of colonies in roll cultures at the end of—	
	One hour.	Four hours.
¾ cubic centimeter=½ per cent. (vol)...................................	None	None.
1½ cubic centimeters=1 per cent............................	... do	Do.
2¼ cubic centimeters=1½ per cent	...... do	Do.
3 cubic centimeters=2 per cent	...do	Do.
Check tube	Countless...	

The table shows that a one-half per cent. solution by volume is sufficient to destroy the bacilli within one hour. What share the sulphuric acid has in determining this result can not be inferred unless tried by itself under precisely the same conditions. This is what has been done in the following experiment:

Five Erlenmeyer flasks, containing each about 150 cubic centimeters of a culture fluid prepared precisely like that in the preceding experiment, were inoculated with hog cholera bacilli and allowed to stand at a temperature of 70° to 80° F., for four days. At the end of this period each drop of the culture medium, as ascertained by roll cultures, contained countless bacteria. Commercial sulphuric acid was added from a sterilized burette to the flasks in definite proportions by volume and roll-cultures made at the end of one, two and a half, twenty-four hours, and four days by transferring a minute quantity of the agitated culture on a platinum loop.

The accompanying table indicates the result obtained, the specific gravity of the acid being 1.8+.

Volume of sulphuric acid added to 150 c. c. culture-liquid.	Per cent. by weight.	Check.	Number of colonies in roll cultures at the end of—			
			One hour.	Two and one-half hours.	Twenty-four hours.	Four days.
c. c. .2	.24	Countless ..	Countless...	Countless ..	100	None.
.4	.48	... do	30	10	None	Do.
.95	1.14	....	None	3	.. do	Do.
1.7	2.04	....	....do	None	.. do	Do.
3	3.6	....do	....do	...do		Do.

One half per cent. of sulphuric acid (by weight) is thus sufficient to sterilize a richly albuminous liquid in three to four hours. Remembering that in the crude carbolic acid mixture the per cent. of sulphuric acid (by weight) contained in the one-half per cent. solution (which was capable of sterilizing a similar liquid in one hour) was .36, we must conclude that there can be no great difference between the crude carbolic acid mixture and the sulphuric acid in regard to disinfecting power. We shall, however, recommend the former in disinfection, since it may last longer in the soil in which sulphuric acid soon forms sulphates and thereby loses its germicide properties. According to experiments given on page 90, .05 per cent. sulphuric acid was sufficient to destroy hog cholera bacilli in ten minutes when no organic matter was present. In the above experiment, in which the liquid contained much organic matter, about ten times as much was necessary.

ORDINARY LIME AS A DISINFECTANT FOR HOG CHOLERA.

Recent experiments made with ordinary lime (CaO) by Liborius (*Zeitschrift f. Hygiene* II, 1887, p. 15) have shown that water containing but .0074 per cent. of lime is capable of destroying typhoid bacilli in the course of a few hours. Cholera spirilla are destroyed by a solution containing .0246 per cent. of lime. These two diseases resemble hog cholera in the mode of dissemination of the virus. In all, the stools are the chief vehicle of the bacteria. Disinfection, therefore, becomes a most important aid in the prevention of the disease.

Lime has many advantages over other disinfectants. It is, first of all, not poisonous. It may be used almost anywhere with impunity where mercuric chloride or strong acids are inapplicable. The soil, when containing the germs of the disease, is not injured by being covered with a small quantity of powdered or slaked lime. The material is exceedingly cheap and can always be obtained without difficulty. The experiments given in the following pages show that lime is a very efficient disinfectant with reference to hog cholera virus, and therefore with reference to the bacteria of swine plague, which are far less resistant than the former.

The method is, in the main, based upon that used by Liborius, with some modifications, which it is not necessary to point out here.

(1) It was desirable to observe what percentage of lime in solution was necessary to destroy hog cholera bacteria not mixed with any appreciable quantity of organic matter. For this purpose water containing different quantities of lime in solution received about two drops from a beef-infusion culture, each drop containing approximately 500,000 bacteria. At the end of one-half hour about one-thirtieth cubic centimeter of the fluid (usually 8 to 10 cubic centimeters) was removed with a flamed platinum spiral and transferred to liquefied gelatine in a test tube. The same was done at the end of three hours. This tube was then placed in ice water and twirled between the fingers while in a horizontal position.

The gelatine, as it congealed, uniformly coated the inside of the test tube, which was immediately transferred to a box in connection with a refrigerator, where it was kept at a temperature of 75° F.* The temperature of the laboratory during these experiments was frequently as high as 95° F., scarcely ever below 80° F., (July). Any bacteria not killed by the disinfectant would show as minute yellowish-white points in the layer of gelatine within forty-eight hours. The number of points corresponded to the number of living bacteria introduced into the gelatine. These cultures will be denominated roll cultures in the succeeding pages. By this method it was found that lime-water diluted with three times the quantity of distilled water was sufficient to destroy hog cholera bacteria in one-half hour. A dilution containing six times the quantity of water destroyed the bacteria in three hours, while a dilution containing twelve times the quantity of water was not capable of destroying them in twenty-four hours.

If we take .12 per cent. as the quantity of lime in lime-water, .03 per cent. of lime will destroy all bacteria in one-half hour, .019 per cent. in three hours.

I.

Lime-water diluted so as to contain of CaO.	Number of colonies in roll cultures prepared at the end of—		Liquid cultures inoculated at the end of twenty-four hours.
	One-half hour.	Four hours.	
0.072 per cent	None	None	Sterile.
0.06 per cent	... do	....do	Do.
0.04 per cent	(*)	....do	Do.
0.03 per cent	None	... do	Do.
Check tube	Countless.		

II.

0.03 per cent	None	None	Sterile.
0.019 per cent	Very many	....do	Do.
0.002 per cent	....do	Very many	Turbid.
Check tube	Countless.		

* Contains hay bacilli.

The quantity of the disinfectant used must be increased, with many disinfectants, with the quantity of organic matter present in the material to be disinfected. The preceding experiment gives us therefore only the minimum quantity that is necessary to destroy hog cholera bacilli when organic matter is practically absent.

A second series of experiments was therefore made by mixing beef infusion cultures with varying proportions of lime-water, and observing

*A description of this method is given by its author, E. Esmarch, in the *Zeitschrift für Hygiene*, i (1886), p. 293.

the quantity necessary to destroy all bacteria present. A preliminary experiment was made by adding lime water in varying proportions to culture tubes containing about 10 cubic centimeters each of beef infusion which had been inoculated the day before and were now opalescent. A flocculent precipitate formed after the addition of the lime-water, which soon settled to the bottom, the quantity of this precipitate varying directly with the amount of lime water added. ·The following table shows that in all the cultures made by transferring a loop of the mixture of culture liquid and lime-water in the proportions there given at the end of thirty minutes, four hours, and twenty-seven hours. no destruction or retardation of bacterial growth could be detected. In the first tube containing beef infusion and lime-water in the ratio of 2 to 1 the precipitate left at first a perfectly clear supernatant liquid. This did not, however, mean complete disinfection. as the roll cultures proved. Moreover, at the end of three or four days this liquid became opalescent again, owing to the rapid unchecked multiplication of the contained bacteria.

A quantity of lime amounting to .04 per cent. was not sufficient to destroy the vitality of beef infusion cultures.

III.

Beef infusion.	Lime water.	Lime.	Number of colonies in roll cultures prepared at the end of—			
			One-half hour.	Four hours.	Twenty-seven hours.	Seven days.
c. c.	c. c	Per cent.				
10	5	.04	Countless	Countless	Countless. ...	Countless.
10	2	.02	...do ...	...do	. do	Do.
10	1	.01	...do	...do	...do	
10	½	.0057	...do ...	...do	...do	
10 *	....	....	...do	....	....	

*Check.

This experiment was therefore continued with relatively larger quantities of lime. Four portions of 40 cubic centimeters each of beef infusion in culture flasks, in which hog cholera bacteria had multiplied for twenty-four hours, were mixed respectively with one-half, one. one and one-half, and two volumes of lime-water, and thoroughly shaken. It was found by testing with roll cultures at the end of thirty minutes, four, and twenty-seven hours that for every 40 cubic centimeters of culture liquid it required 60 cubic centimeters of lime-water to destroy all bacteria within twenty-seven hours, while a large number were destroyed within four hours. It required 80 cubic centimeters to completely sterilize the culture fluid in four hours, although nearly all bacteria were destroyed within the first half hour. In the flask to which 40 cubic centimeters had been added there was a partial destruction of bacteria at the end of twenty-seven hours. At the end of six days the bacteria were as numerous as in the flask containing 20 cubic centimeters of lime-water, in which no destruction of bacteria or retardation of growth could be observed. The percentage of lime in these cultures is given in the following table:

IV.

Beef infusion.	Lime-water.	Lime.	Number of colonies in roll cultures prepared at the end of—			
			One-half hour.	Four hours.	Twenty-seven hours.	Six days.
c. c.	c. c.	Per cent.				
40	20	.04	Countless.	Countless.	Countless.	
40	40	.06	....do......	....do.....	Very many.	Countless.
40	60	.072	....do......	A few....	None.....	None.
40	80	.08	A few....	None......	....do....	

When the lime-water was added to the beef infusion a light flocculent precipitate was formed immediately as in the preceding experiment, the quantity depending on the amount of lime-water added. This settled to the bottom in a very short time, leaving a perfectly clear, slightly yellowish layer of liquid above. In those flasks, however, in which disinfection was not accomplished, this layer of liquid remained turbid, or else became clear at first, with the partial destruction or precipitation of bacteria, and then became clouded again as the remaining bacteria multiplied in it.

The quantity of organic matter which is present in substances to be disinfected is ordinarily quite large. This is true of bowel discharges which contain the specific hog cholera bacteria, and which should therefore be thoroughly disinfected. To prepare a solution containing a considerable amount of insoluble albuminous matter, the method of Liborius was adopted. Beef was chopped finely and allowed to soak over night in the refrigerator in twice its weight of water, as in the preparation of beef infusion. After the meat had been removed from the liquid in a press the latter was neutralized and the white of an egg was added in the ratio of one egg to every 600 cubic centimeters of liquid. The whole was boiled and the coagulated masses allowed to remain in the liquid. The amount of solid particles and lumps, when deposited on the bottom, formed a layer nearly as deep as that of the liquid above it.

This mass of liquid and solid matter was placed in Erlenmeyer flasks, each receiving 150 cubic centimeters. The whole was sterilized for several hours at a steam pressure of 12 pounds and subsequently inoculated with hog cholera bacteria. At the end of three days, when each drop of the culture liquid contained nearly a million of bacteria, milk of lime, made by adding lime in the ratio of one gram to 9 cubic centimeters of

water, or. in other words. milk of lime containing 10 per cent. of lime,
was added in various quantities as shown by the following table:

v.

No.	Beef Infusion.	Milk of Lime.	Lime.	Number of colonies in roll cultures made at the end of—			
				One-half hour.	Three hours.	Twenty-four hours.	Six days.
	c. c.	c. c.	Per cent.				
1	150	40	2.1	None	None*	None ...	
2	150	20	1.2	....do	....do	...do ...	
3	150	10	.62	... dot ...	....do	.. do ...	
4	150	5	.32	....dot	....do	...do ..	None.
5	150	2	.13	Very many	Very many.	50	Very many.
	(¿)	....	....	Countless.			

*About 12 fungi.   †About 100 fungi.   ¿Check (beef infusion).

In all the experiments the cultures containing the precipitated matter were shaken up before testing so that the quantity taken for the
roll cultures (about $\frac{1}{20}$ to $\frac{1}{30}$ cubic centimeter on a spiral of platinum)
consisted of solid particles as well as liquid.

Within thirty minutes after the addition of the milk of lime the precipitate had settled so as to leave the supernatant liquid perfectly
clear in flasks 1, 2, and 3; in flasks 4 and 5 it was still turbid as before the addition of the lime. It cleared up in No. 4 soon after. but remained permanently turbid in No. 5. In none of the other flasks did
this layer become turbid even after seven days. In these the bacteria
were permanently destroyed, as the table shows. That the lime transferred to the roll cultures could have had no retarding effect on the
growth of any bacilli present was proved by adding at least four times
the quantity of lime to a gelatine tube which had been inoculated with
a drop of culture fluid. Countless colonies appeared in the gelatine
layer in due time. The check tubes were tested only at the beginning
of every experiment to test the vitality of the cultures used. Since hog
cholera bacteria will remain alive in the beef infusion employed for
weeks and months, there was nothing to be gained in making roll cultures from these check tubes more than once.

The table shows that 5 cubic centimeters of a 10 per cent. milk of
lime was sufficient to sterilize within one-half hour 150 cubic centimeters beef infusion containing much suspended albuminous matter. This
would be equivalent to about one-half gram of ordinary unslaked lime.
or about .0032 gram for every cubic centimeter of liquid, i. e., .32 per
cent. We may assume, therefore, that for every twenty pounds of
fecal matter or discharges from diseased pigs only one ounce of lime
in the form of milk of lime is needed, provided the two are thoroughly

mixed. The less complete this mixing can be the more lime must be added to make the disinfection thorough. By comparing the amount of lime necessary when there is little or no albuminous or other organic matter present and when there is a large quantity of it, we are impressed with the importance of always taking into consideration the circumstances under which disinfection is to take place. Thus the bacteria of hog cholera are destroyed in one-half hour when placed in water containing only .03 per cent. of lime. When there is present beef infusion diluted to one-third the original strength it requires .08 per cent. of lime. When the suspended albuminous matter is considerable it requires .32 per cent.

It may be more convenient to use unslaked lime either in small lumps or powdered shortly before application to the mass to be disinfected. That this may be done the following experiments are sufficient proof:

Erlenmeyer flasks, containing each from 100 to 200 cubic centimeters of beef infusion in which there was a large amount of suspended matter, as in the preceding experiment, were used for this purpose. After inoculating each with hog cholera bacteria they were allowed to stand for forty-eight hours. At the end of this time the flasks contained many millions of germs in each cubic centimeter. Unslaked lime was broken up into lumps as large as peas or beans, and thoroughly heated over a Bunsen flame to destroy any adherent spores and drive away any moisture. After cooling, this was thrown into the culture-flasks in the proportion, by weight, of 2, 1, $\frac{1}{2}$, and $\frac{1}{4}$ per cent., respectively. The flasks became very slightly warmer. After thorough shaking the suspended matter soon began to settle down, leaving a clear, supernatant liquid, which remained clear after ten days' observation. The liquid in the last flask ($\frac{1}{4}$ per cent. lime) cleared up most tardily. The attached table shows that even as little lime as $\frac{1}{2}$ gram in 100 cubic centimeters was sufficient to permanently destroy all bacteria in a liquid very rich in albuminous substances. Even at the end of eight days the flasks containing the smallest amounts of lime were sterile.

VI.

Beef infusion.	Per cent. of CaO.	Number of colonies developed in roll culture prepared with $\frac{1}{10}$ cubic centimeter of infusion at the end of—			
		One hour.	Four hours.	Twenty-four hours.	Eight days.
100 cubic centimeters plus 2 grams CaO.	2	None	None	None	
120 cubic centimeters plus 1.2 grams CaO	1	....do	A few......	do	
200 cubic centimeters plus 1 gram CaO.	.5	Very many..	None	.. do	None.
200 cubic centimeters plus .5 gram CaO.	.25	....do	.. do	.. do	Do.

The table shows that in the flasks containing one-half and one-fourth per cent. of lime there were still a considerable number of living bacteria at the end of the first hour, although destroyed at the end of four hours. The presence of bacteria in the flask containing 1 per cent. of lime at the end of four hours, although none were detected at the end of the first hour, shows that all bacteria are not necessarily destroyed simultaneously.

A second experiment, conducted precisely as the preceding, confirmed the latter in every respect. The quantities were chosen somewhat differently, as the appended table shows, in order to find out the lowest percentage of lime that will destroy all bacteria within a given period of time.

VII.

Beef infusion.	Lime.	Number of colonies in roll cultures made with $\frac{1}{20}$ cubic centimeter of the beef infusion at the end of—			
		One hour.	Three hours.	Twenty-four hours.	Five days.
	Per cent.				
100 cubic centimeters plus 1 gram CaO .. ,	1	Very many .	None	None ..	None.
200 cubic centimeters plus 1 gram CaO .	½	None	....do	...do	Do.
200 cubic centimeters plus .5 gram CaO ...	¼	Very many .		50 ...do	Do.
200 cubic centimeters plus .3 gram CaO ...	.15	Countless...	Very many..	50	Do.
Check tube		...............do			

The lime was powdered, and heated in a platinum evaporator to destroy all adherent spores. There was scarcely any perceptible rise of temperature when the lime was added to the liquid. The first flask contained a culture eight days old, the remainder contained cultures four days old. From the table it will be seen that one-half per cent. of unslaked lime is sufficient to sterilize a very turbid albuminous liquid in four hours, one-fourth per cent. in twenty-four hours. Fifteen-hundredths per cent. was almost enough to destroy all the germs in one day while at the end of five days all were dead.

DISINFECTION OF THE SOIL WITH LIME.

The same fine loam used in the experiments to determine the vitality of hog cholera bacteria in the soil was used here. The method pursued was briefly as follows:

Into small beakers, plugged with cotton wool and sterilized at 150° C. for several hours, about 50 grams of slightly moist soil was introduced, and the whole sterilized under steam pressure at 110° C. About 10 cubic centimeters of a beef-infusion culture of hog cholera bacteria was then stirred up with it. After a certain length of time milk of lime was added and thoroughly mixed with the soil. The destruction of the bacteria was noted at certain intervals of time by taking small bits of the soil on a platinum loop and making roll cultures therefrom.

The table appended gives the results of a series of experiments carried out according to this plan:

Number and date of experiments.	Quantity of lime in soil.	Check culture	Number of colonies in roll cultures prepared from a small bit of soil after—							
			One-half to one hour.	Three to four hours.	One day.	Two days.	Three days.	Five days.	Six days.	Seven days.
1888.	Per cent.									
I.—Jan. 18	1	[1] ∞	[2]	0	0	0	....	.....	......	
II.—Jan. 20	2	∞	100	[3] 0			....	.....	......	
III.—Jan. 23.	½	∞	[4] 0	0	3–4	0	10	.....	......	
IV.—Jan. 24	¼	∞	[2]	250	1000	∞	....	.....	......	
V.—Jan. 31	½	∞	[2]	.....	20	0	....	[2]	[2]	
VI.—Jan. 31	¼	∞	∞	[2]	8	.....	∞	....	∞	∞
VII.—Feb. 10	½	[2]	....	0	0	.....	0	0	......	0
VIII.—Feb. 10	¾	[2]	0	0	0	.....	0	0	......	0
IX.—Feb. 23 [5]	½	∞	[2]	0	0	.....	[6]	.....	......	
X.—Feb. 23 [5]	¾	∞	[4]	12	0	.....	[6]	.....	......	
XI.—June 7 [7] ...	¾	∞	[2]	[2]	0	0	....	0	0	0
XII.—June 7 [7].	1	∞	0	0	0	0	....	0	0	0

[1] Countless colonies.
[2] Less than ∞, but still too numerous too be estimated.
[3] Fungi present.
[4] A few fungi present.
[5] Rabbit died of hog cholera after inoculation with soil on nineteenth day.
[6] A micrococcus multiplies in the soil.
[7] Rabbit inoculated with soil on eleventh day remains alive.

The following may serve to explain more fully the tabulated experiments and results obtained:

I. Fifty grams sterile moist soil, infected with 10 cubic centimeters of a beef-infusion culture January 5. January 18: 5 cubic centimeters of a 10 per cent. milk of lime (9 cubic centimeters water and 1 gram ordinary unslaked lime) was stirred up with it. This is equivalent to 1 per cent. of lime. A roll culture from the infected soil was made in all experiments before the lime was added, to make sure of the presence of living germs. Roll cultures, to which three *loops* of milk of lime were added, showed no diminution in the growth of colonies, proving that the small amount of lime (fraction of a loop) added with the soil had no retarding effect. In this experiment 1 per cent. lime was sufficient to remove from the soil countless bacteria (as shown by check-roll) in three to four hours.

II. January 20: To another beaker of soil, prepared and infected with the preceding, 2 per cent. milk of lime was added and stirred up. The result identical with Experiment I.

III. January 23: 2½ cubic centimeters of 10 per cent. milk of lime added (= ½ per cent.). The check-tube contains so many colonies as to have an opalescent appearance. Disinfection is nearly completed after one hour. A few colonies develop in the one day and three day tube.

IV. January 24: 1¼ cubic centimeters of 10 per cent. milk of lime (= ¼ per cent.); no disinfection is brought about. During the first twenty-four hours there is a decided diminution in the number of colonies, but a decided increase thereafter.

V, VI. January 31: The sterile soil in two beakers was infected yesterday with a beef-infusion peptone culture three days old, 10 cubic centimeters being added to each beaker. To-day enough milk of lime

(5 per cent.) is added to two beakers to make one-half per cent. and one-quarter per cent., respectively. The check tubes show subsequently countless colonies. The table shows a slight diminution in the number of colonies for the first twenty-four hours, then an increase. No disinfection. For the one-half per cent. beaker there is apparently a complete destruction during the first two days, but subsequently the few remaining multiplied again. No disinfection here. This result seems to disagree with the third experiment. The discrepancy was, however, caused by the fact that the milk of lime was not thoroughly shaken up when used.

VII, VIII. In these experiments the lime was thoroughly stirred up before use, and in both beakers disinfection has taken place. The germs in the soil before adding the lime were not so numerous, because the soil had been infected from a liquid culture only fifteen minutes previous and no multiplication could have taken place.

IX, X. The conditions of the experiment are the same as before. The soil is infected with a liquid culture two days old. Seven days later check-tubes are prepared from each beaker of soil. One receives one-half per cent. lime, the other three-quarters per cent. (*i. e.*, 5 cubic centimeters and 7.5 cubic centimeters of a 5 per cent. milk of lime to 50 grams soil). The table shows that the beakers were invaded by a micrococcus after the first day, which multiplied enormously in the soil, so that hog cholera colonies could not be detected. After nineteen days two rabbits, inoculated with a little infusion made from the soil, succumbed to hog cholera.

March 13: Rabbit inoculated subcutaneously with sterile beef infusion, in which a little of the one-half per cent. lime soil had been stirred up. Rabbit dead March 19. Spleen enlarged, congested; kidneys, lungs, and duodenum with hemorrhagic foci. Beginning necrosis in liver.

A second rabbit inoculated from the three-quarters per cent. lime soil in the same way and at the same time. Rabbit dead March 21. Lesions the same, in addition to a hemorrhagic condition of the lower portion of large intestine. In both hog cholera bacteria present.

XI, XII. Experiments carried out as the preceding ones. The bacteria were permanently destroyed, as shown by the roll cultures made up to the seventh day, and unsuccessful inoculation of rabbits later on the eleventh day.

From these experiments we may conclude that three-quarters to one per cent. of lime will destroy hog cholera bacilli in the soil. It is highly probable that a smaller per cent. of lime (one-quarter to one-half per cent.) will be amply sufficient when simply scattered in a thin layer over the surface where the great majority of the disease germs will remain until destroyed by natural agencies.

IS THERE ANY RESISTANT SPORE STATE IN THE LIFE HISTORY OF THE BACILLUS OF HOG CHOLERA?

This question can now be answered with the help of the foregoing experiments.

Stained in dilute aqueous solutions of aniline colors the bacilli from the tissues of animals which have succumbed to the disease stain in such a way as to leave the impression that each bacillus contains an endospore. A narrow band of stained substance bounds an oval pale body, which is but slightly tinged. It appears that a rather resistant envelope prevents the coloring matter from passing readily into the interior of the bacilli.

If a drop from a recent liquid culture be suspended from the lower surface of a cover-glass and examined in a glass cell with a homogeneous immersion objective and small diaphragm, the following appearances are worthy of record: The bacteria in the center of the drop of culture fluid are in very active motion. If the periphery of the drop be examined there will be found a dense layer of bacteria caught there by the slow desiccation and consequent contraction of the drop. These, some of which are still moving slowly, appear slightly larger than the forms in the center of the drop.. As the drying proceeds and the film of water becomes thin, the bacteria appear to be made up of a distinct dark border surrounding an almost transparent body. In most forms there is a slightly thicker border at the ends than at the sides of the short, rod-like bodies. When stained slightly this border takes the stain well, while the body of the rod remains pale. The fact that the structural and color pictures correspond is strong evidence that the microbe possesses a rather dense membrane, which in optical section is seen as a narrow dark border.

Involution forms may occasionally simulate endogenous spore formation. When hog cholera bacilli are placed upon the surface of gelatine with a drop or two of blood, a few days are sufficient for the formation of a large number of filaments from two to many times the length of the bacilli as they are found in the tissues of animals. The ends are rounded or irregularly pointed; the width of the bacilli varies; in general, they have the physiognomy of abnormal forms. In each bacillus there may be from one to four oval, square, or oblong, clear spaces. When stained they remain without color, thus simulating spores. When carefully examined in a fresh condition, there is no refrangibility—they are mere holes or spaces left by the irregular breaking up and retraction of the protoplasm.

Microscopical characters, however, are now and then misleading, unless we interpret them by physiological experiments. Judging from what have hitherto been considered properties of bacterial spores, the microbe of hog cholera can not lay any claim to the production of true endogenous spores. Their absence is determined by results of experiments recorded in the preceding pages:

(1) The thermal death-point of the bacilli at 58° C. An exposure to this temperature for fifteen to twenty minutes destroys not only the vitality of cultures of all ages, but also the germ in the tissues of the infected animal. A momentary exposure to boiling water is equally efficacious.

(2) The bacteria are destroyed by disinfectants in solutions which are incapable of destroying spores.

(3) They are killed by simple drying far more quickly than are spores; at the same time their resistance to drying is much greater than might be expected under the circumstances. In the experiments recorded some dried bacteria in spleen pulp were killed in less than a month; others resisted forty-nine days. For cultures we may put the limit, according to experimental data, between one and four months. It is this continued vitality in the dried state that suggests the existence of a membrane which is more resistant than that possessed by the great majority of bacteria in their vegetative state.

(4) Subjected to various conditions of moisture and dryness, of freezing and thawing in the superficial layers of the soil, they are destroyed after an exposure of between two and four months.

The facts brought out by the study of the bacillus lead to the conclusion that a distinct spore state, so called, does not appear either within the animal body or in nature.

WAYS IN WHICH SWINE BECOME INFECTED.

BY WAY OF THE DIGESTIVE TRACT.

(*a*) *Feeding diseased viscera.*—In at least 90 per cent. of swine. hog cholera may be induced by feeding to them the viscera of animals which have died of the disease. The lesions produced are exceedingly severe. The mucous membrane of the large intestine is extensively ulcerated or completely necrosed. In animals which have contracted the disease in the ordinary way in infected pens the ulceration of the large intestine, at times very severe, usually stops abruptly at the ileo cæcal valve. When this is slit up, the mucosa belonging to the small intestine up to the free border of the valve is in the great majority of cases normal, while the mucosa of that surface of the valve facing the cæcum may be extensively ulcerated. In many animals fed with infectious matter the ulceration involves the entire ileum. This is well illustrated by the following cases:

January 8, 1886.—Pig No. 165 was fed with the viscera of two pigs which had died of hog cholera. It was found dead January 26, after manifesting no marked symptoms of disease except a tendency to lie quietly in its pen. On examination the subcutaneous fat was found diffusely reddened. There was a slight peritonitis, indicated by a considerable quantity of straw-colored effusion and some fibrinous stringy deposits. There were also a few local excrescences on the small intestine, due to the irritation of *echinorhynchi*. Spleen somewhat enlarged; on its surface a few bright red punctiform elevations. Right heart distended with a clot. Local hepatizations in lungs, probably caused by lung worms, which were very numerous. Stomach but slightly reddened. A number of ulcers in the duodenum, the mucosa of which was reddened. The mucosa of the ileum for 1½ feet from valve was completely necrosed, the walls thickened. and the serosa of this portion dotted with ecchymoses. On the upper portion of the ileum there were scattered ulcerations on a deeply congested membrane for 6 or 7 feet. The entire length of the large intestine was covered with dirty, yellowish ulcerations varying in diameter from a pin's head to nearly an inch. The mucosa itself was very deeply congested in the cæcum and colon, and the walls much thickened. *Ascarides* and *echinorhynchi* numerous in small intestines. The liver attached to diaphragm in several places by a whitish exudate.

A tube of meat infusion with peptone inoculated from the spleen of this animal was found to be a pure culture of the motile bacillus of hog cholera. Line cultures on gelatine plates confirmed the microscopic examination. A tube of nutritive gelatine inoculated from the spleen at the same time contained in each needle-track, several days later, from ten to fifteen colonies of the same organism. Two cover-glass preparations revealed no bacteria. This fact, combined with the small number of colonies in the tube culture, gave evidence of the small number of germs in the spleen tissue. Inoculations on mice and guinea-pigs gave substantially the same results as those obtained hitherto.

No. 159 was fed with viscera of No. 165 on January 28. February 5, its eyes were sore and nearly closed; it was quite weak. It died the following day, only eight days after infection. The skin on abdomen was reddened in patches; the subcutaneous tissue diffusely. The superficial inguinals, as well as the glands in the abdomen, were deeply congested, the cortex more especially. Those of the thorax were nearly pale. The spleen was dotted with a few punctiform blood-red elevations. Beneath the epicardium and endocardium of both auricles and the endocardium of the left ventricle were extensive patches of extravasated blood. Kidneys enlarged and congested throughout. The lesions of the ileum, caecum, and colon in this animal were quite as extensive as those of the case just described; there were no ulcers in the rectum, however. Those of the colon had black centers, pointing to a recent origin from blood extravasations on the surface of the mucous membrane.

In the spleen of this case the characteristic bacteria of hog cholera were exceedingly numerous, as determined by cover-glass preparations. Two liquid cultures proved pure when tested on gelatine plates. In the needle tracks of a tube culture in gelatine innumerable colonies appeared in a few days. Inoculations into animals from subsequent cultures proved equally positive.

Pig No. 156 was fed with the viscera of No. 159 February 18, and after manifesting the usual symptoms of hog cholera, died February 25, seven days after feeding. Among the marked lesions produced by the disease was a complete necrosis of the upper two thirds of the colon, with scattered ulcers along the lower third. Eight feet of the lower portion of the small intestine, beginning at the valve, was necrosed. In the spleen there were numerous small grayish spots, probably centers of necrosis, as they showed no longer cell structure when crushed on a slide and stained. The fundus of the stomach was also deeply congested.

The spleen, to which organ the microscopic examination was limited, contained the characteristic bacteria, as shown by cover-glass preparations. Three liquid cultures made from the same organ were found to be pure cultures of the same microbe when tested by line cultures. A tube culture in gelatine developed in each needle-track numerous non-lique-fying colonies.

In these animals the mode of introduction of the virus determined the seat of the severest lesions. It is probable that the food passes quite rapidly through the small intestine; that in the stomach the action of the bacteria is more or less limited, because they have not sufficient time to multiply, and probably because hindered by the acid condition of the organ, though they will multiply with considerable vigor in slightly acid solutions. The prolonged stay of the food in the large intestine permits multiplication, and thereby causes the first and severest lesions to appear here. When these have become very extensive, so as to paralyze the action of the large intestine, the ileum becomes involved in a similar manner, possibly by a partial stoppage of the infectious matter in this portion of the intestine.

It may therefore be said, in general, that the feeding of hog cholera viscera produces lesions like those found in natural infection, only more severe and more extensive. The duration of the malady is also much shortened. The notes given above are quoted from one of the reports of the Bureau. Since they were written numerous other pigs have been fed in this way in the course of vaccination and other experiments with precisely the same result. We must therefore emphatically deny the truth of statements made now and then in agricultural journals that hog cholera can not be communicated to healthy swine feeding upon the viscera of those dead from this disease. Such statements are pernicious in tendency, and aggravate the evil which this malady carries with it. *

(b) *Feeding pure cultures of hog cholera bacilli.*—The successful reproduction of this disease by feeding pure cultures of hog cholera bacteria proves not only that the bacteria fed are the true cause of the disease, but also that infection may and does take place in this way. We again quote from the report for 1886 some experiments which demonstrate that swine may take the disease and die by simply swallowing somewhat more than half a pint of beef broth in which hog cholera bacteria were growing.

December 13, 1886.—Three pigs were fed with 300 cubic centimeters each (three-fifths of a pint) of a beef-infusion culture of hog cholera bacilli kept in the thermostat, at 95° F., for three days. The culture was contained in two flasks. When examined both were found pure. The pigs were prepared for the feeding as follows: No. 348 received no food for over twenty-four hours. A 2 per cent. solution of sodium carbonate in beef infusion was then given to increase the alkalinity of the stomach. Of this about 1 liter was consumed. It was then fed with 300 cubic centimeters of culture liquid mixed with beef broth to make 1 liter. No. 350 was starved in the same way, but received no alkali before consuming the culture. No. 342 was not deprived of food before eating the culture.

The result confirmed our anticipations. No. 348 showed signs of disease in two days. On the third it was unable to rise, and died on the same day. The *post-mortem* examination showed a considerable congestion of the mucous membrane of the duodenum and jejunum, as well as of the large intestine. The fundus of stomach affected in the same way. The liver was gorged with blood, as well as the portal system. There were no marked lesions of the other viscera. That hog cholera bacilli had also entered the blood was shown by two pure cultures in beef infusion obtained from the spleen. A gelatine culture from the liver contained about six or seven colonies.

No. 350 was a more typical case, and demonstrated the severe local effects of the bacillus much better, since the animal lived longer. It ate fairly well until the fourth day, when its appetite gave way and diarrhea set in. From this time it grew weak and thin, being scarcely able to walk. It died on the tenth day after feeding. The lesions of the alimentary tract were exceedingly grave. Beginning with the stomach, the mucous membrane was dotted with closely-set elevated masses as large as split peas, and larger patches of a whitish viscid substance, made up entirely of cellular elements (diphtheritic?). When removed, a raw, depressed surface was exposed. The membrane itself

---

* It may be possible that those who make this claim were confronted with swine plague. This disease, although resembling hog cholera very closely in many features, we have not been able to reproduce by feeding.

was pale. Besides a general injection of the ileum, Peyer's patches were more deeply congested, and the uppermost covered with a thin, yellowish film, not removable, and most likely dead epithelium. In the cæcum and colon the mucosa was superficially necrosed, and converted into a continuous layer of a dirty whitish mass about 1ᵐᵐ thick. The walls of the intestine were greatly thickened and very friable.

Microscopic sections showed an extensive cellular infiltration of the submucous connective tissue which had separated the masses of fat cells, concealed the connective tissue fibers, and caused a great thickening of the entire layer. The mucosa itself was greatly altered. The surface was necrosed and converted into an amorphous mass. In some places the necrosis involved the entire depth of the crypts of Lieberkühn, a series of striæ indicating their former existence. Those whose epithelium still remained were plugged with a cylindrical mass, inclosing broken-down nuclei. The bacteria had exerted their poisonous effects from the surface of the mucosa towards the depths, destroying the surface epithelium and glandular structures and involving secondarily the submucous layer. Near the rectum this continuous mass of dead tissue was replaced by isolated ulcers embedded in an intensely reddened mucosa. The ileo-cæcal valve was much swollen, but the necrosis did not extend into the ileum, although there were a few ulcers near the valve, and the epithelium had a pale, lusterless aspect, as if dead. The liver was filled with blood, which readily clotted as it flowed from the cut surface. Spleen congested and but slightly enlarged. Lungs hypostatic. The lymphatic glands in general not much affected. Two liquid cultures from the blood were turbid next day, and contained hog cholera bacilli only. In a gelatine tube culture from the liver about a dozen colonies developed in each needle-track.

No. 342, which was fed with the same quantity of culture liquid, but was not deprived of food previously, was somewhat ill on the following day. It recovered, however, and continued apparently well for several weeks. It began thereupon to grow thin and weak. On January 26 it was no longer able to rise, and was therefore killed for examination, in order to conclude the experiment. On opening the abdominal cavity it was at once perceived that the animal had been suffering from a very intense disease of the large intestine, a portion of which was firmly attached to the bladder. When dissected out and slit open, the mucous membrane of the cæcum and colon was found replaced by a brownish friable layer of necrosed tissue. The wall of the intestine was infiltrated to such an extent that it was nearly one-fourth inch thick, and so degenerated that the forceps easily tore through it. The thickness of the walls prevented the intestine from collapsing after it was opened. Its only contents was a brownish liquid mass. The glands of the meso-colon were very large, some like horse-chestnuts. On section the entire tissue was very pale, almost white. The spleen was somewhat enlarged; the Malpighian corpuscles unusually large and prominent on section. Lungs and heart normal; kidneys deeply reddened throughout.

This case is very interesting in completing the information gained by this feeding experiment. No. 348, which had been fed with sodium carbonate, besides being deprived of food, died three days after the ingestion of the culture. No. 350, which was simply starved, died ten days thereafter, while No. 342, which ate the culture without being previously starved, was dying on the thirty-fourth day.

These results show that infection may occur by way of the digestive system, provided the destructive action of gastric digestion be prevented, as was done by starving and by the use of an alkaline carbonate.

They also indicate how purely local this destructive action may be. Gelatine cultures from these animals showed that the internal organs contained but very few bacteria. So few were they in fact that the microscope alone could not have demonstrated their presence in the spleen.

In addition to the cases given, equally positive results were obtained at four different times by the feeding of pure cultures. Some of these results have been mentioned in other places (see page 207 of the Report of Bureau of Animal Industry for 1885). It was also found that small quantities up to 100 cubic centimeters ($\frac{1}{5}$ pint) may be fed to most pigs without obtaining any fatal results. Sometimes the fed animal becomes very sick, but recovers; sometimes no disturbance whatever is produced.

These experiments prove conclusively the causal relation between the bacilli of hog cholera and the disease so called. They show that pigs fed with nothing but sterile beef infusion, to which the minutest speck of growth from a hog cholera culture was added and in which the bacilli thus introduced were allowed to multiply for one or more days, were destroyed by a disease identical with hog cholera, but far more rapid in its course and more severe in its manifestations.

SUBCUTANEOUS INOCULATION.

Subcutaneous inoculation with hog cholera bacilli from cultures is successful in only a small percentage of cases, except when these germs are unusually virulent. In the report for 1885 several cases of successful inoculation are given, among which the following deserve to be briefly quoted:

November 27, 1885.—Two pigs (Nos. 112, 114) were inoculated subcutaneously into the thigh with 3 cubic centimeters each of a pure liquid culture. No. 114 died nine days after inoculation. The superficial inguinal glands were swollen, with hemorrhagic points in medulla. Spleen enlarged, dark. Extravasations on auricular appendages of heart. Lungs œdematous; bronchial glands enlarged, dark red throughout (hemorrhagic). Glands of abdomen in general hemorrhagic, except those of mesentery; petechiæ under serosa of cæcum. Kidneys with glomeruli appearing as blood-red points, the entire organ congested. Mucosa of fundus of stomach, lowest portion of ileum, and of the cæcum and colon deeply reddened with slight extravasation. Cover-glass preparations as well as cultivations in gelatine and beef infusion revealed hog cholera bacilli, and these only.

No. 112 died on the 15th day. Diarrhea appeared two days before death. The lesions in this animal resembled those of No. 114, with the following differences: Spleen very large, dark, friable. Kidneys less congested. Lungs with minute hemorrhages throughout the parenchyma. Ecchymoses beneath the endocardium of left ventricle. Lymphatics and digestive tract even more congested and hemorrhagic than in No. 114.

The success of these inoculations among numerou, failures in subsequent trials must be ascribed to the exceptionally virulent disease of that year, for no cases have been observed since which died so suddenly and presented such severe hemorrhagic lesions of the various vital organs.

The following experiments, made with blood taken from the heart of swine affected with hog cholera and killed for the purpose, are taken from the report for 1886:

September 10.—A pig dying with the disease was killed, the heart carefully exposed, and the blood drawn with a disinfected hypodermic syringe. Nos. 329 and 333 received subcutaneously 5 cubic centimeters each, one-half in each thigh. No. 329 in a few days lost its appetite, became weak and stupid. Found dead October 5. Slight local swelling at the points of inoculation; superficial inguinals greatly enlarged; hypostatic congestion of lungs; complete necrosis of mucous membrane in cæcum; large scattered ulcers in colon, showing as whitish patches on serous surface and encircled by a crown of enlarged blood vessels; bacteria in spleen.

No. 333. Slightly ill for a time; fully recovered; died December 2, with no other lesions than engorgement of liver; no signs of former ulceration.

A second experiment was made in the same way.

October 13.—Nos. 324 and 325 inoculated as in the preceding experiment, 10 cubic centimeters of blood being used for each animal. No. 324 was found dead November 1, after being off feed for a time; deeply reddened skin over caudal half of abdomen; superficial lymphatics extremely large and serously infiltrated; on section, hemorrhagic points; at point of inoculation the connective tissue is infiltrated; 50 to 75 cubic centimeters clear amber serum in abdominal cavity; papillæ of kidneys deeply reddened; slight congestion, but no ulceration in large intestine; lymphatics in general moderately tumefied and congested.

No. 325 found dead October 29; reddening of skin as in 324; extravasation in connective tissue; spleen greatly enlarged, purplish; lymphatics of thorax and abdomen purplish—enlarged; petecchiæ on section of kidney and pelvis, also over entire surface of epicardium; lung tissue mottled both on surface and on section with purple spots, due to blood extravasation into alveoli, so that it scarcely floats; mucous and serous surface of small intestine dotted with petecchiæ; small hemorrhages on the surface of the mucous membrane and into the submucous tissue of the cæcum and upper colon; ulceration beginning.

On first thought we might be inclined to attribute these successful results to a greater virulence of the germs in the injected blood. This view needs further confirmation, however. The injected blood coagulating in the connective tissue contains in it the bacteria, which are not only protected from the aggression of cellular elements, but have actually a store of nourishment upon which they may live and multiply. No such advantages are presented to bacteria suspended in liquids which are readily absorbed, leaving them to the mercy of the tissues surrounding them. The local reaction in the above animals was very insignificant compared with that produced by liquid cultures. In order to come to any conclusion it would be desirable to add a few bacteria from cultures to fresh blood, and observe the relative virulence by subcutaneous inoculation.

Elsewhere in this volume will be found a small number of successful inoculations with pure liquid cultures among a large number of unsuccessful ones. The inoculations were made for the purpose of determining whether they would confer immunity. In reading over these experiments it will also be seen that such injections produce swellings varying from the size of a pea to a hen's egg. The tumor is developed in the subcutis and consists of a yellowish white, tough tissue, breaking down in the center into a grumous mass after a period of one or more months.

In view of these facts we can not consider wounds or bites inflicted on the surface of the body a means of infection, excepting, perhaps, in epizootics of extreme virulence. Nor can we ascribe much influence to the stings of insects in inoculating the virus from one animal to another, as this would correspond to subcutaneous inoculations on a minute scale. Flies may, however, infect food in various ways, to be dwelt upon in subsequent pages.

INTRA-VENOUS INOCULATION.

The injection of hog-cholera virus (from the spleen of a pig examined near the city of Baltimore, Md.) into the circulation was tried but once, and this trial was successful in producing a hemorrhagic septicæmia such as we frequently encounter in outbreaks of the disease, and which proved fatal in less than three days.

November 12, 1888.—Pig No. 90, black and white, about five months old. The right crural vein was exposed by raising a triangular flap of skin over it after thoroughly disinfecting the latter with a one-fifth per cent. solution of mercuric chloride. Five cubic centimeters of a beef infusion peptone culture inoculated from an agar culture about a week old was injected into the exposed vein with a hypodermic syringe thoroughly disinfected with 5 per cent. carbolic acid. The liquid culture was two days old when used. Two hours after the inoculation the temperature had risen from 103⅗ F. to 107°. November 13 there was no swelling, but a slight serous oozing at the place of inoculation. The appetite was good. November 14, at 3 p. m., the temperature was 107⅘. The animal was disinclined to move, although it came to eat in the morning and evening. November 15 it lay on its side quietly, with occasional kicking. Found dead at 4 p. m. Autopsy held immediately.

General blush on skin of ventral aspect, snout, and lips. No swelling at the point of inoculation; slight blood extravasation. Spleen enormously enlarged, 14 inches long, 2 to 3 inches wide, and one-half to 1 inch thick, gorged with dark blood, and friable. Superficial inguinals enlarged, œdematous; on section diffuse pale red spots: cortex congested. Bronchial and renal glands enlarged, partly hemorrhagic, gastric glands hemorrhagic throughout substance. The blood is thick, dark colored, coagulation slight, even after several hours' exposure to the air. Several petecchiæ on epicardium of right auricle. Right side of heart distended with blood: in it a small white clot. Left side contracted, empty. Lungs normal, excepting one-third of left ventral lobe, which is collapsed. Kidneys enlarged, deeply congested throughout. The surface is thickly dotted with minute deep red points. The papillæ so deeply reddened that any extravasations would be unrecognizable. A few petecchiæ in pelvis. Bladder contains about 30 grams of urine tinged with blood. The whole mucosa of stomach is deeply congested. In fundus it is hemorrhagic with numerous patches of necrosed epithelium one-fourth to one-half inch across.

The upper 8 inches of duodenum in the same condition as the fundus of stomach. Numerous red points scattered over mucosa of entire small intestine. In lower ileum a few hemorrhagic points. The mucosa of cæcum and upper colon very slightly congested, but the remaining two-thirds intensely so. Hemorrhage here and there sufficient to stain the feces with blood which were otherwise normal. The mesenteric and meso-colic glands all deeply congested throughout their substance.

Cover-glass preparations from spleen pulp showed a large number of hog cholera bacteria. Cultures from the same revealed the presence of the same organisms only.

INFECTION BY WAY OF THE LUNGS.

It has been maintained by some observers that infection may take place through the lungs. Bacteria dried and carried about with the dust are supposed to be deposited in the lung tissue by the inspired air. It is of considerable importance to determine whether germs carried by currents of wind can in this way produce an outbreak at a distance. There are two questions involved which must be dealt with separately in order to get as clear an idea as possible of this rather perplexing subject. (1) Do bacteria of hog cholera when inhaled produce a disease which is similar to the disease described in these pages, i. e., without extensive lung lesions; and (2) do hog cholera bacteria produce a specific disease of the lungs when there deposited just as they produce intestinal disease when swallowed? The first question, whether the lungs may be considered an entrance of the virus when in the dried condition, is one for which not sufficient proof is at present forthcoming. Several experiments have been made at the experiment station to produce the disease by spraying pigs in a tight box with liquids containing hog cholera bacteria and by injecting them suspended in liquids into the trachea. In no instance was the disease produced. Several times bacteria were injected through the walls of the thorax into the pleural cavity and lung tissue without any result, excepting in two instances given on page 55. In one animal there was some pleuritis but no pneumonia. On the other hand the large intestine in this animal was severely diseased as in natural infection. In the other animal death occurred *seven* months after inoculation. The lung disease was most likely the combined effect of lung worms and the mechanical injury resulting from the inoculation.

If we must conclude, therefore, that animals are but rarely infected through healthy lungs, it is still less likely that these bacilli produce quite uniformly disease of the lungs, as has been assumed by other observers without adequate proof. In the notes of the outbreak given on page 54, this subject has been pretty thoroughly discussed in the light of evidence furnished by bacteriological tests. The results there reached may be briefly summarized that lung disease or pneumonia of a severe character is not often found in hog cholera. The slight broncho-pneumonia frequently met with is most probably due to catarrhal conditions and to the aspiration of foreign bodies. There is, however, this very diseased condition of the lungs which may favor infection through them. When there are present foci of broncho-pneumonia, hog cholera bacteria inhaled may lodge there, multiply in the secretion of the smaller bronchi, and cause an extension of the disease not possible in healthy lungs. From there they may be coughed up and swallowed, carried into the large intestine, where they exert their most destructive activity. In this sense the lungs may become the entrance of the virus, but only when previously diseased. In the examination of such diseased lobules of the lungs, in the outbreak referred to, hog cholera bacteria were found in almost every case. In these cases it is difficult to decide whether they entered the diseased lobule from without by way of the bronchi— i. e., with the air inspired—or whether they were deposited there by the circulating blood.

The question of the relation of lung disease to hog cholera has been complicated by the existence of an infectious pneumonia in swine, which we have called swine plague. This disease, which will be treated of in a special publication, appears over a large part of our country at times in virulent epizootics. It is caused by a widely distributed septic organism, and appears occasionally in swine suffering from hog cholera. In a number of such cases both disease germs have been obtained from the same animal.

We do not exclude infection through the lungs as improbable in hog cholera, especially in epizootics characterized by more than usual virulence. At the same time long experience at the experiment station has shown that swine do not become infected in pens only a few hundred feet away from pens full of sick and dying animals. Currents of air can thus have but little power in distributing germs capable of inducing the disease.

SOME OBSERVATIONS ON THE PATHOLOGICAL ACTION OF HOG-CHOLERA BACTERIA.

We have frequently recurred to the fact that the intestinal tract seems to suffer more or less exclusively in hog cholera. The changes there produced belong chiefly to the necrotic and ulcerative type, combined with a variable amount of neoplastic growth in the bottom of the ulcer. In connection with this work it has been impossible to make any extended observations on the genesis of these ulcers owing to the exactions of other work. Whatever may be suggested here must be regarded simply in the light of inferences from lesions as they were observed *post mortem*.

The ulcers are produced in most cases by a process beginning at the surface of the mucosa. This is shown very well by sections of ulcers in which only a portion of the tubular glands have been necrosed. What this process is must be left an open question. It seems very probable that the bacilli invade the tubules and blood-vessels of the mucosa, by their rapid multiplication plug the latter, and cause a coagulation necrosis of the most superficial portion of the membrane. The various bacteria present in the intestine then complete the breaking down of the membrane, and an ulcer is formed. Ulcers may, however, be formed by bacteria carried thither by the blood, for ulcers are present in those animals upon which subcutaneous inoculation has been practiced (p. 110). In these cases the plugs or infectious emboli are situated probably in the submucosa. The resulting ulcer is deeper and more extensive.

The formation of plugs or thrombi is confirmed by inoculations in rabbits (p. 69). The process of coagulation necrosis, as the result of plugs in the capillaries of the liver, may be seen in almost every case. Microscopic examination likewise reveals their growth in masses in the internal organs.

The nature of the hemorrhagic lesions so frequently observed in hog cholera demands some attention. The rupture of minute and sometimes larger blood vessels is no doubt due to the same process of thrombosis and embolism in which the thrombi and emboli may be made up more or less entirely of bacteria. Whether these plugs have a directly destructive action (necrosis) upon the delicate vascular wall so as to produce rupture and extravasation, as has been suggested in former reports, or whether the action is merely mechanical, or whether both causes are at work, it would be impossible to say. At all events these hemorrhages are frequently the precursors of ulcers in the intestine, and in some cases the only lesions which were observed were hemorrhages from large and small vessels. We may picture to ourselves the process, beginning, in most cases, with local ulceration in the large intestine (cæcum). The bacilli may be carried thence through injured vessels into other parts of the body, and to other portions of the intestine, where fresh ulcers may appear; or the bacilli may be carried onward from the first ulcers to portions of the intestine lower down, where they again attack the mucous membrane.

In many cases the quantity of virus introduced into the digestive tract is so great—as when pure cultures of hog cholera bacilli are fed—that destruction of the mucous membrane goes on at the same time throughout the cæcum and colon, becoming less severe as the rectum is approached. The paralysis of the large intestine may cause extensive necrosis of the ileum. In these cases the action of the bacilli is purely superficial at first, penetrating deeper and finally reducing the entire wall to a friable mass, greatly swollen by the infiltration of leucocytes. In such feeding experiments we may also observe grades of lesions as we go away from the valve in the ileum and in the colon. Instead of the conversion of the mucosa into a thickened, discolored, and roughened membrane, an amorphous, dirty grayish exudate may be found on it. The destruction, going on very slowly, causes a reactive inflammation, manifested by the exudate.

The genesis of the intestinal lesions in the rabbit is somewhat different. They appear quite late in the disease as a result of the discharge of bacilli from the necrotic foci of the liver tissue into the bile duct. They are limited to the duodenum and large intestine, and are especially severe in the latter, forming there peculiar hæmatomata or blood tumors, accompanied with a fibrinous exudate. The same lesions are obtained by feeding cultures. In view of this localization in the rabbit and pig we must admit that hog-cholera bacilli exert their primary action where they are retained for a considerable time by the feces, and thus have an opportunity of multiplying before they attack the mucous membrane.

The size of the spleen, sometimes enormous, seems to stand in no direct relation to the number of bacilli present. Sometimes a very highly congested spleen harbors only the average number of germs.

The presence of hog cholera bacilli in the lung lesions met with in a small percentage of cases has already been dwelt upon (p. 54). The lesions themselves show nothing characteristic. They are usually restricted to the most dependent lobe, the ventral on one or both sides, and to the writer they simply indicate a proneness of catarrhal inflammation in young pigs, due, no doubt, to the aspiration of food and dirt, consequent occlusion of bronchi, atelectasis, and broncho pneumonia. The hog cholera bacilli which may be found here in most cases are present for the same reason that they are found in the spleen. Circulating in the blood, they find the contents of the diseased and plugged air tubes and alveoli a very suitable medium for multiplication. In this way they may add to the irritation and cause extension of the disease.

It seems pretty certain that the chief damage done by hog cholera bacilli lies in their power to grow in plugs in the small vessels. The simple reason why we have such severe lesions in the intestinal tract, although the bacilli are present in small numbers in the internal organs also, lies in the anatomical nature of mucous membranes in general and their exposed situation with reference to putrefactive processes.

If the obscure action of these bacilli on cells and cell-life could be formulated at all, in the present state of our knowledge the writer would assume: 1. That they exercise a more mechanical influence by thrombosis, which leads, in the mucous membranes, to circumscribed necrosis. Death occurs not by the direct poisonous action of ptomaines, but from secondary causes, such as changes in the liver, septic infection. 2. In acute cases the bacilli possess a greater power of invading the circulation. There they multiply rapidly enough to cause rupture of blood vessels in all the vital organs. Death may occur in these cases by a ptomaine poisoning of the nerve centers due to the large number of bacilli in the blood, or the loss of so much blood drawn from the circulation, or, what is more probable, by hemorrhages in the substance of the central nervous system. Whether this greater ability to invade the vital organs depends upon the production of a more poisonous ptomaine, which paralyzes opposing forces, or to other physiological properties, belongs to the future for solution.

Concerning the variation in the destructive nature of hog cholera bacilli little that is positive can be said. We have observed, for instance, that the disease dies out after a time in a herd, very frequently because there are no more animals left, or because the few that have survived have been exposed in such a way as to acquire immunity. But it has also been observed at the experiment station that the disease may disappear, even when fresh animals are exposed to it from time to time. The decline in virulence may be due to the particular season with its frosts or droughts, and other meteorological changes, or it may be due to a gradual weakening of the virus as it passes through the system of the exposed swine. We have observed several times that when a series of animals are fed with infected viscera, each from the one preceding it, the earlier ones of the series will develop the most severe disease. The later ones are affected by a more chronic malady, and finally a point is reached when no disease is produced. In such an experiment self-inoculation or vaccination must be eliminated, as the animals are not exposed previous to the feeding. But aside from this gradual dying out of particular epizootics, we have noticed a difference in the outbreaks themselves from year to year. In some the acute, hemorrhagic type would prevail, in others a more chronic ulcerative type. Such differences are most likely due to modifications of the disease germ itself, although here also it is difficult to eliminate differences in the animals due to race, feeding, cleanliness, etc., and differences due to the time of the year. As regards the manifestation of differences in the bacteria themselves none can be formulated.

BACTERIOLOGICAL INVESTIGATIONS OF HOG CHOLERA IN NEBRASKA, ILLINOIS, AND MARYLAND.

The facts which have been gathered concerning the bacillus of hog cholera were obtained chiefly from epizootics occurring in the District of Columbia. The advantages of a bacteriological laboratory connected with the Bureau made investigations possible which could not be made in the field. It was, however, necessary to confirm the results already obtained by investigations elsewhere. These have led to the discovery of the same bacillus in Nebraska and subsequently in Illinois, thus identifying the disease in widely-separated localities.

Nebraska.—In March, 1886, the bacillus of hog cholera was isolated from one of ten spleens sent from Kansas and Nebraska. Though carefully removed by the inspector and placed in sterilized bottles plugged with cotton wool, none came in good condition, owing to the long delay on the road. Small bits of spleen were placed under the skin of a number of mice. Two which had been inoculated from the same spleen died of hog cholera. These experiments are reported in full in the Third Annual Report of the Bureau, and therefore need only to be summarized in this place. The bacillus obtained in a pure condition from these mice was carefully tested upon other mice, upon rabbits, and upon pigeons. The pathogenic effect was the same as that of the bacillus from eastern outbreaks. It was, however, readily observed that the Nebraska bacillus was less virulent, as it did not kill guinea pigs. A number of inoculation and feeding experiments were made upon pigs in order to produce the disease. All but two failed to show any disease. These two, fed with liquid cultures, were made ill, and one very sick was killed on the ninth day after feeding. The large intestine showed intense congestion, with a few patches of a diptheritic membrane in the cæcum. No bacteria were obtained in cultures from the internal organs. This experiment, which was not very satisfactory, was repeated by feeding a larger quantity of culture liquid. The result was entirely satisfactory, as the following description of the experiment will show. Moreover, the feeding took place nearly nine months after the microbe had been obtained originally from the spleen. During this time it had been cultivated in various media, passed through mice and rabbits, so that no trace of any spleen substance or foreign material could have been present in the culture liquid which was fed. The result also indicates that the pathogenic power of the bacillus was in great part retained during this time.

A flask containing between 500 and 600 cubic centimeters of sterile beef infusion was inoculated from a culture (rabbit), and after standing six days in the incubator the entire amount was given to a pig which had not been fed for thirty six hours. The culture liquid was covered with a thin membrane, and on microscopic examination contained only the motile bacillus. The animal became dull and weak, eating little. The bowels were loose on the fourth day. On the fifth it was unable to rise, and on the sixth it was found dead. The autopsy notes are briefly as follows:

In the abdominal cavity several hundred cubic centimeters of a reddish serum, a thin translucent exudate covering the peritoneum of the intestine, which is diffusely reddened. Between the layers of the mesentery, along the line of attachment to small intestine, an abundant translucent gelatinous exudate. Spleen very dark on section; the surface dotted with elevated points of extravasated blood. Liver congested. Lungs normal, with exception of a few lobules, which are simply collapsed.

Almost the entire digestive tract was found involved. Around the cardiac orifice of the stomach a zone of mucous membrane about 2 inches in width was covered with whitish diphtheritic patches. Isolated ulcers in duodenum. About 6 or 7 feet of the lower portion of the small intestine very much thickened, the mucous membrane covered with a thin sheet of necrosed tissue, whitish, brittle. The cæcum and portion of the colon greatly thickened and covered with a thick layer of necrosed tissue very rough and brownish. Near rectum necrosis gives way to closely set, isolated, roundish diphtheritic elevations of a whitish color, which leave a raw surface when scraped away.

These lesions were, therefore, as intense as any produced by feeding pure cultures of hog cholera bacteria obtained from the East. The specific identity of the two bacteria from Nebraska and the East was thus completely established.

In the liver and spleen the bacteria were few, for a cover-glass preparation from each organ did not show any after some searching. Liquid cultures from the blood (heart), spleen, and liver were turbid on the following day, and all contained the motile bacillus. Within four days complete membranes had formed on the surface. That the bacteria were very few in blood from the heart was indicated by a gelatine tube which had been inoculated several times with a platinum wire dipped in blood; no colonies were visible on the fourth day. Of three liquid cultures, each inoculated with a loop of blood, two remained sterile. It was presumed that the liver would contain the largest number, inasmuch as the portal circulation received its blood from the seat of the disease. This assumption was confirmed by the very abundant colonies surrounding a piece of liver tissue which had been dropped into a tube of nutrient gelatine.

In order to make certain of the pathogenic powers of the cultures obtained from this case of feeding the tube culture from the liver was used to infect two rabbits. The skin on the inner aspect of one thigh was carefully shorn and disinfected with .1 per cent. corrosive sublimate. An incision was made through the skin, and with a loop dipped in the surface growth of the culture a minute quantity was introduced into this pocket. The larger of the two rabbits was found dead on the fifth day. The lesions were briefly as follows:

Slight amount of pus at the place of inoculation. Neighboring inguinal glands enlarged and infiltrated with blood. Surrounding vessels much injected and very tortuous. Liver very friable, spleen dark and enlarged; both dotted with points and stellate spots of coagulation necrosis, especially numerous on the caudal surface of the liver. Both organs contained the bacteria of hog cholera in large numbers. Lungs hypostatic. A small number of ecchymoses beneath pleura; very few bacteria in kidneys and heart's blood.

The second rabbit was found dead on the sixth day. The lesions were the same, if we except the more pronounced coagulation-necrosis in the liver and its absence in the spleen. The bacteria of hog cholera were distributed as above; very abundant in spleen and liver; lungs normal.

Gelatine tube cultures from spleen and liver of these rabbits confirmed the microscopic examination. Liquid cultures from the blood contained the motile bacteria, which had formed a brittle surface membrane on the second day.

Differential characters of the hog cholera bacillus from Nebraska.—The bacillus when stained on cover-glass preparations from the spleen and other viscera, closely resembles the one found in the disease prevalent in the East, so that it is impossible to distinguish them in this way.

A few minor differences revealed in the various culture media indicated, however, that the two microbes were not alike in every way, and brought up the very interesting question of the variation of species of bacteria and the influence of such variation on the severity of epidemics.

The most noteworthy difference was observed in liquid cultures. Within twenty-four hours after inoculation from the spleen or blood the culture liquid became turbid, and upon its surface a complete membrane was present in nearly every case. This whitish membrane is not homogeneous, but made up of patches of varying thickness, and when shaken slowly settles to the bottom in lumps and flocculi. The microbe of Eastern outbreaks does not form a membrane within several days after inoculation, and then only when the tube remains perfectly quiet. It appears as a whitish ring attached to the glass.

In other respects the two microbes differed so little that no mention need be made of these differences here. We have, therefore, only the membrane on liquid cultures and a slightly diminished virulence, by which the bacilli from Nebraska were distinguishable from those obtained from Eastern epizootics. Both of these characters are variable ones. The membrane can be produced in part by cultivating the Eastern germ for a long time at a high temperature. The variation in virulence is observable during the course of almost every epizootic.

Illinois.—From the spleen of an animal affected with hog cholera the specific bacillus was obtained during some investigations conducted in this State in the summer of 1886. The bacillus presented no differnces whatever from the one obtained in Eastern outbreaks. Its effect upon rabbits and mice was precisely the same. Its fatal effects when fed to a pig is well illustrated by the following experiment:

A pig was kept without food for over twenty-four hours. A 2½ per cent. solution of sodium carbonate in meat broth was then given to it. Of this it consumed about one liter, taking thus about 25 grams of the salt. It was then fed with about 50 cubic centimeters of gelatine cultures and 100 cubic centimeters of liquid cultures of the hog cholera bacillus obtained from rabbits which had succumbed to inoculation and from the original gelatine culture of the spleen made in Champaign County, Ill. The animal was found dead December 4, scarcely three days after feeding. This was the briefest period of illness thus far observed. The lesions were very pronounced. Pyramids of kidneys deep red throughout; glomeruli visible as dark points. Lungs pale, not fully collapsed. Right heart filled with semi-coagulated blood. Liver gorged with blood. Mucosa of stomach intensely reddened, especially along fundus, and covered with a thick layer of tenacious mucus. Mucous membrane of ileum similarly affected. Peyer's patches exceedingly dark red, showing through serous coat. When viewed from the mucous surface the elevated border gives each a slightly concave boat-shaped appearance. The colon also deeply congested, almost hemorrhagic in patches, filled with a small quantity of semi-liquid feces. The rectum still filled with consistent masses. The mesenteric glands congested.

The feeding had thus produced a very severe inflammation of the digestive tract. The diagnosis was further confirmed by obtaining pure liquid cultures from the spleen, the liver, and blood from the heart. The bacteria were not sufficiently numerous in these organs to be detected by the microscope. To make sure that the carbonate of soda had no corrosive effect, another animal was treated precisely in the same way by starving and feeding a solution of the salt. No ill effects whatever were manifested.

In order to test the specific pathogenic character of the bacillus obtained from this animal a large rabbit was inoculated subcutaneously with about one-eight cubic centimeters of a liquid culture from the blood. On the sixth day the rabbit was lying on its side; abdominal breathing very labored. It was found dead on the next day. Slight thickening of the subcutaneous tissue and fascia covering the thigh muscles at the point of inoculation. The muscular tissue covered with minute ecchymoses around the infiltrated patch. Small quantity of serum in the peritoneal cavity. Spleen very large, blackish, exceedingly friable, and crowded with bacteria. Liver enlarged; interlobular tissue pale; the entire parenchyma very soft and brittle. Dotting both surfaces of all the lobes are small, grayish-white patches, involving one, two, or three, rarely more, acini, and bounded very sharply by the acini themselves. Peculiar figures are thus formed, three contiguous ones giving the patch a clover-leaf appearance. On section they are found to extend to the depth of one or several acini into the parenchyma. The great majority of these masses of coagulation-necrosis involve lobules on or near the surface. Only a few are in the depths of the organ. When such a whitish mass is spread on a cover-glass and stained innumerable bacteria of hog cholera make their appearance. The rest of the tissue is likewise crowded with them. Beneath the pulmonary pleura are large purplish patches of extravasation, which on section extend deeply into the parenchyma. The lung tissue is in general congested. Blood from the heart contained very few bacteria. No cultures were made.

Maryland.—During September and October, 1888, an epizootic of hog cholera was started at the experiment station by a sick pig brought from near the city of Baltimore, Md. The hog cholera bacillus had been previously obtained from the spleens of two animals killed at Baltimore. At least twenty, which had been either simply exposed in pens or fed with the viscera of dead animals, succumbed to the disease. In all cases pure cultures of the specific bacillus were obtained directly from the spleen. Its effect upon rabbits and mice, its appearance and mode of growth, did not differ from the bacillus as described in the foregoing pages.

RELATION OF HOG CHOLERA TO THE PUBLIC HEALTH.

The importance of knowing whether or not certain infectious animal diseases have any deleterious influence upon human health, or are the direct cause of human disease, is not even second to the economic importance attaching to animal diseases. There are some animal diseases, such as glanders, tuberculosis, anthrax, and rabies, which are directly communicable to man. This transmissibility was known before the specific bacterial organisms of these diseases had been discovered. There are other animal diseases not proved to be transmissible to man, whose causes have only been definitely recognized within the past few years. As regards these it is quite pertinent to ask, in the light of present knowledge, whether they have any relation to human diseases. Animal diseases when communicated to man may become so changed in character as to be unrecognizable. Hence we must determine whether the specific micro organisms of animal diseases are found in human diseases or not; in other words, whether the causes are identical. During the course of these investigations this problem has been constantly borne in mind. The question to be settled first has reference to the presence of hog cholera bacilli in human diseases. There are two maladies of mankind which resemble hog cholera in many respects—typhoid fever and dysentery.

Typhoid fever is an infectious disease of man, prevalent throughout the civilized world, while hog cholera is known to exist only in North America and the British Isles, and during last year it has been identified on the continent of Europe. Hence there could be no close relationship between the two diseases, for they would naturally occur together if they were identical. But there is another reason for their non-identity; the micro-organisms which produce them are different. Thus the typhoid fever bacillus is a longer rod than the hog cholera bacillus, its movement in liquids is more sluggish and of a somewhat different character. Its growth on boiled potato is almost peculiar to itself. Moreover, it has no pathogenic effect upon any animal thus far tried, and we know that the hog cholera bacillus is fatal to mice, rabbits, guinea pigs, and pigeons after subcutaneous inoculation, and to pigs when fed to them. Hence there remains not the shadow of a doubt that genuine typhoid fever and hog cholera are distinct diseases.

They do, however, belong to the same general class of infectious diseases. In both, infection very likely occurs chiefly through the food and drink. In both the lesions are ulcerative in character, located in the pig in the upper portion of the large intestine, in man in the lower portion of the small intestine. It is true the ulceration may be the secondary stage of processes primarily different anatomically speaking, but we know as yet too little of the relative effect of irritants of the same kind, but of different intensity, to venture any positive statement. In both diseases the invasion of internal organs by the specific micro-organisms takes place, so that they are readily obtained from the spleen. In both they appear in clumps or colonies in the capillaries. There is also a general resemblance of the two organisms, both as regards their appearance, motility and growth in various media when we compare them with the bacteria of other infectious diseases in man and animals. We may therefore conclude that although these two diseases are entirely distinct there exists a close relationship between them as to the cause, the manner of infection, and the type of disease.

Hog cholera is allied to the disease or group of diseases known as dysentery even more closely than to typhoid fever. It resembles, perhaps, most nearly that form called epidemic dysentery. Of the causes of this disease very little is known. That the epidemic type is caused by micro-organisms of some kind is now generally accepted, and it is also believed that there may be several distinct species of micro-organisms which are capable of producing diphtheritic and ulcerative lesions of the large intestine.* Anatomically hog cholera and diphtheritic dysentery are very much alike. The question whether hog cholera virus can ever produce a dysenteric affection in man is not answered until we have learnt more of the causes of the latter disease.

The danger, if any there be, of causing disease in man would only occur upon farms where the opportunity is afforded for infection of food and water by the discharge of diseased pigs. The preparation of pork by cooking is sufficient to destroy the bacilli if by any chance the flesh of diseased animals should get into the market. Experiments have shown that a temperature of 140° F. is sufficient for their certain destruction. The temperature reached in boiling, roasting, etc., is of course much higher than this.

In addition to these conclusions, which follow from the facts developed by our experiments, it may be said that with the thousands of *post-mortem* examinations made by ourselves and other investigators, there has not been a single case where the operator became infected. In several instances it has been recorded that fresh wounds upon the hands have been covered with virulent material during the examinations without producing any appreciable effect. And through all the years that this disease has prevailed so extensively in the United States there has not been a case reported where it has been shown on good evidence to have been transmitted to the human species.

Notwithstanding these facts, however, we should expect, with the prevailing carelessness in disposing of the carcasses of swine in many sections during such outbreaks, that there would be a more or less deleterious effect upon human health, not from the specific nature of the disease, but from the decomposing organic matter. Too often these carcasses are left to putrefy near dwellings, or are thrown into streams to contaminate them with decomposition products. Such practices are dangerous to the health of the community and should be prevented by the local sanitary officers.

---

* Investigations of the Egyptian dysentery by Koch and Kartulis have shown that the cause is an organism belonging not to the class of bacteria, but to the protozoa, and assuming the form of an amœba. This is found in large numbers in the ulcerated walls of the intestine.

PREVENTION OF HOG CHOLERA.

It is frequently necessary to apply preventive measures before infectious diseases have actually appeared in a herd. The disease may have appeared on a neighboring farm, and the problem then arises : How can the infection be prevented from spreading to other farms ? How can the surrounding farms keep the malady from their premises ?

The sources and channels of infection are as follows, the most common and important being placed first :

(*a*) *Pigs purchased from infected herds, or coming in contact with those from infected farms, or running over grounds occupied by diseased swine within two or three months.*

(*b*) *Infected streams may communicate the disease to herds below the source of infection.*

(*c*) *Virus may be carried in feed, implements, and on the feet and clothing of persons from infected herds and premises.*

(*d*) *Winds, insects, birds (particularly buzzards), and various animals may transport hog cholera virus.*

(*a*) In regard to *a*, it may be said that no pigs should be purchased from any locality until one year after the death of the last case of cholera. There are frequently near the end of an epizootic chronic cases which may live for three or four months without showing any distinctive signs of disease until they suddenly die. The *post-mortem* examination usually reveals extensive ulceration of the large intestine. The disease may thus linger in a herd long after all danger has apparently subsided. By bringing any chronic cases in contact with hitherto unexposed healthy swine the disease may spring up anew, as a dying fire would when supplied with fresh fuel. Although our experiments have shown that the disease germs may all disappear from the soil in three or four months, the uncertainty of knowing whether there are any chronic cases continually adding fresh virus to the soil makes the period of one year not too long to avoid the introduction of unexposed pigs. It is advisable, in districts where hog cholera is very prevalent and is rarely absent for any length of time, for farmers to raise their own pigs and not trust to any animals from outside. In this way infection may be at least in part kept under control. When animals have been obtained from places which are not above suspicion they should not be brought in contact with swine already on the place, but quarantined as far as possible from them and kept under careful observation for at least one month.

(b) Perhaps the most potent agents in the distribution of hog cholera are streams. They may become infected with the specific germs when sick animals are permitted to go into them or when dead animals or any part of them are thrown into the water. They may even multiply when the water is contaminated with fecal discharges or other organic matter. Experiments in the laboratory (p. 76) have demonstrated that hog cholera bacilli may remain alive in water for four months. Making all due allowance for external influences and competition with other bacteria in natural waters, we are forced to assume that they may live at least a month in streams. This would be time enough to infect every herd along its course.

(c) Hog cholera germs are not immediately destroyed by drying. Laboratory experiments show that they may retain their vitality from two to four months. Hence it is not difficult to see how a person walking on infected ground and among infected animals may carry on his shoes and clothing dried germs of the disease to any neighboring herd. For the same reason hog cholera germs may be carried from infected grounds to others by feed, and by farming implements which have come in contact with infected ground.

(d) There is no reason to suppose that currents of air have much influence in spreading the disease. Observations at the experiment station of the Bureau have left no doubt that healthy pigs may be kept on the same farm with diseased ones without becoming infected, provided the infection is not carried in feed and implements, or on the shoes and clothing of persons from the sick to the healthy. Moreover the disease is an intestinal malady, and all evidence points to infection through the food rather than through the air inspired.

The agency of flies and other insects is perhaps equally limited when infection is to be carried from one place to another. Our experiments show very well that the sting or bite of an insect is not sufficient to produce the disease. It is possible, however, that they may carry the virus from one place to another in the same yard. This will be discussed more fully under another head.

The agency of buzzards in distributing the disease in the Southern States seems probable, although there is no positive proof. These birds will readily consume carcasses of dead swine. If the hog cholera germs are not destroyed by digestion it is reasonable to assume that the feces contain the living germs, which may cause the disease to break out at some distant place. Of course, the remedy would be to immediately destroy or bury dead animals.

There is some reason to believe that rats, dogs, and perhaps other small animals may carry the germs upon their feet or in their hair and thus infect premises. It is probable that the contagion is only rarely transported in this manner, but there are outbreaks the origin of which it is difficult to explain otherwise. We may readily conceive, bearing in mind the facts enumerated above, how such animals might become contaminated with moist, semi-liquid, or liquid matters containing the germs, and that these substances drying upon the feet or hair would adhere for a considerable time. If an animal thus infected should go into uninfected lots occupied by swine and deposit there the smallest particle of the germ-containing material, either in liquids standing in the feeding troughs or in moist organic matters suitable for the multiplication of the microbes, an outbreak of the greatest virulence might be set up. The number of conditions which must coincide for the spread of the disease in this way probably accounts for the comparative rareness of this kind of infection.

Granted, then, no communication between infected and uninfected farms, there still remains the danger of infected water-courses, upon which it is impossible to lay too much emphasis. In fact, if the disease exists anywhere along a stream all farms below that point are liable to infection unless use of the water in any form whatever is given up during the season.

By paying particular attention to these points, there is no doubt that the disease can be warded off, even when in the immediate neighborhood. Hog cholera is analogous to typhoid fever, dysentery, and Asiatic cholera in man in many particulars, and there is a quite unanimous opinion that these diseases are most commonly transmitted through drinking water. The same may be predicated of hog cholera, and the mysterious spread of this plague will no doubt frequently be understood by examining the water courses.

When the disease is in the neighborhood it has been customary with some to feed swine on some so-called "preventive" medicine. These are frequently prepared or invented by individuals who have little, if any, knowledge of the action of medicines. The outcome is that the animals fed with these unknown compounds are not only not benefited, but their vitality is actually reduced, and when the disease appears it destroys the weakened animals much more easily. The writer has made *post-mortem* examinations of several animals in the West where such preliminary treatment was going on, and the peculiar changes of the internal organs, not like any known disease, could only be referred to the action of such preparations. It must be remembered that there are few medicines which are not injurious or poisonous in large doses. They should not be used thus excepting under special conditions, and only given as recommended by those who have been trained to know the peculiar value and effect of drugs.

The condition of the animals themselves is of great importance in favoring or preventing infection. When pigs are fed with liquids in which the specific bacilli only are present, those that have been deprived of food for some time previous take the disease, while those whose stomachs contain food that is undergoing digestion do not take it readily. If, besides starving the animals, they are fed with some alkaline solution, by which the alkalinity of the stomach is increased, the pathogenic effect is still more pronounced. Any disorder of digestion by which the secretion of gastric juice is diminished or checked and the mucus is increased in quantity will increase the susceptibility of the animal to infection, because the alkalinity of the mucus will favor rather than destroy the virus. Any mode of feeding which produces constipation and over-distension of the large intestine is likely to favor the disease, as the virus is retained for a longer time. During epizootics, therefore, besides the preventive measures suggested, the animals should be carefully fed upon food which tends to keep the bowels open and the feces soft, and which does not interfere with normal digestion.

When there is a suspicion that a herd has been infected, although the disease has not yet appeared, disinfection, and all the rules laid down below (p. 130) should be carried out with great care, as if the disease were actually present.

When hog cholera has appeared in a herd or on a farm, precautions should be taken for two reasons: (1) To prevent the virus from being carried to other farms and infect other herds. (2) To prevent the loss of the entire herd, or, if this is not possible, to stamp out the disease in such a way that the ground shall not infect healthy animals subsequently.

The rules under the first head should be prescribed by law to protect property from the consequences of the carelessness or the willfulness of those who refuse to take proper precautions. They may be summarized as follows:

(*a*) *The dead animals should be immediately disposed of, either by burial or by burning, or if they are taken to some rendering establishment their transportation should be governed by well-defined rules which will prevent the dissemination of virus on roads, in wagons, cars, etc.*

(*b*) *Streams should be carefully protected from pollution.*

(*c*) *No animals should be removed from any infected herd or locality to another free from the disease for at least six months after the last case of disease.*

(*a*) The proper disposal of dead animals is a matter of great importance, for the bodies not only contain the germs of the disease, but the latter will multiply enormously during summer heat in the internal organs after life has been extinguished. Each dead body must therefore be regarded as a focus of the disease unless properly disposed of. It may be buried. In such case it must be so deep that no animal can get at it. It should be covered by a layer of powdered or slaked lime several inches thick, and the ground over the body likewise sprinkled with a thin layer of the same. If the carcasses are burnt, care should be taken that any parts not consumed are buried as directed. If they are carried away some distance to rendering establishments, at best a dangerous pocedure, employés of such establishments should be compelled to wrap around the carcasses impervious cloths wetted with a 2 per cent. solution of carbolic acid. so as to protect the roads from the virulent drippings.

(b) The danger from infected streams has already been mentioned at length. These must be protected by law in such a way that no sick animals should be allowed to go near them, and that no carcasses be thrown into them or deposited where drainage may carry the virus from the body into the water. Nor should the drainage from pens be permitted to flow into them.

(c) Hogs are frequently affected with cholera of a mild form, which lasts for several months before some form of septic infection or degenerative changes in the internal organs produce death ; hence it is important to insist upon knowing when the last case of disease occurred. Since it has been demonstrated that hog cholera germs may remain alive in the soil from three to four months, this rule will not appear unreasonable as a safeguard.

These rules will be sufficient, if properly executed, to confine the disease within narrow limits. There is no doubt that hog cholera virus dies out over the greater part of our country after epizootics have swept over it. We have no reason to believe that it can survive in the soil from one end of the year to the other. It is, in fact, highly probable that it is transported and distributed from a few places where, for some reason, cases have occurred throughout the year and have thus kept the virus alive. There are no experiments on record which show that the hog cholera germ may be found in the soil and water independent of the disease. It has been looked for, but has never been found, excepting in the body or discharges of diseased swine.

In view of the fact that the disease can be kept under control, the legislatures of those States which suffer most severely from this plague should take steps to enact rules similar to those formulated above. The States of Kansas and Nebraska have on their statute-books laws of this character, which read as follows:

AN ACT to prevent the spread of disease among swine.

Be it enacted by the legislature of the State of Kansas. It is hereby made the duty of every person who owns or who has the control of any hog that has died of any disease to bury or burn the same within twenty-four hours after such hog has died, and any person who knowingly fails or refuses to comply with the provisions of this section shall be deemed guilty of a misdemeanor, and upon conviction thereof shall be fined not exceeding one hundred dollars.

SEC. 2. Whoever shall knowingly barter or sell any hog afflicted with any disease without giving full information concerning said disease shall be deemed guilty of a misdemeanor, and upon conviction thereof shall be fined not exceeding one hundred dollars.

SEC. 3. Whoever shall knowingly barter or sell any hog which has died of any disease shall be deemed guilty of a misdemeanor, and upon conviction thereof shall be fined not exceeding one hundred dollars.

SEC. 4. Whoever shall throw or deposit a dead hog in any river, stream, creek, or ravine shall be deemed guilty of a misdemeanor, and upon conviction thereof shall be fined not exceeding one hundred dollars.

AN ACT to prevent the spread of hog cholera and other kindred diseases, and prevent traffic in animals dying from infectious or other diseases.

Be it enacted by the legislature of the State of Nebraska, That it shall be the duty of the owners of swine or other domestic animals dying from cholera or other diseases, within twenty-four hours after their death, to cause the carcasses of such animals to be suitably buried or burned up on the premises owned or occupied by such person.

SEC. 2. If the owner of any swine or other domestic animals dying from cholera or other disease, or any other person, shall sell or dispose of the carcass of such swine or other domestic animals, to any person for the purpose of manufacturing the same into soap or rendering the same into lard, or for other purposes, or if any person shall buy or otherwise obtain the carcass of any swine or other domestic animals, dying from cholera or other disease, for manufacturing purposes as aforesaid, or any other purpose except that of burial or burning, as provided in the preceding section, every such person shall upon conviction be fined in any sum not less than twenty-five dollars nor more than one hundred dollars, or be imprisoned not exceeding six months.

SEC. 3. Whereas an emergency exists, this act shall be in force and take effect from and after its passage.

Approved, March 4, 1885.

These laws, although not sufficiently explicit, touch upon the most important points, and are important movements in the right direction. We would suggest laws embodying the three heads in their entirety as given above under *a*, *b*, and *c*, together with directions for their proper execution. The disease, spreading so easily and rapidly, requires great promptness of action and quite different rules from those which must be adopted in the suppression of glanders or tuberculosis, for example. The difference is due to the nature of the specific microbe, so unlike those causing the two diseases mentioned.

It is not strange that so little attention has been paid to the restriction of this disease in the past, since legislators and boards of health and State veterinarians have had no scientific basis upon which to frame laws. Even now efforts are being made in various quarters to controvert or openly deny the accuracy of the investigations and results obtained by the Bureau, and throw the whole subject back into the chaos in which it was but a few years ago. This must have any thing but a salutary effect upon those intrusted with the framing and execution of specific laws for the protection of domesticated animals.

Having thus far dwelt upon the means which must be adopted to prevent the spread of the disease from one place to another, it becomes necessary to consider some of the measures that should be employed in checking it after it has once taken foothold in a herd. But how are we to recognize the disease? To answer this question it may be well to recapitulate briefly some of the more important features of the malady in as simple language as possible:

It is quite common for the disease to announce itself by a few sudden deaths. The stricken animals may seem well a day, perhaps only a few hours, before death. In order to remove any doubts as to the precise nature of the disease, it is best to examine one or more of the animals before burying or burning them. This should be done in a secluded place which pigs can not reach, and the ground thoroughly disinfected, as will be described later. The disease in the sudden cases can be easily recognized. The spleen is, as a rule, very black and enlarged. Spots of blood from the size of a pin's head to a quarter inch or more will be seen in the fat under the skin, on the intestines, lungs, heart, and kidneys. The lymphatic glands are purplish instead of a pale pink. When the large intestines are opened they are found covered with these dark spots of blood more or less uniformly and entirely. Often the contents are covered with clotted blood. Any or all of these may be considered as signs of the disease in its most virulent form. In many outbreaks the early cases do not succumb so rapidly. They grow weaker, lie down much of the time, eat but little, and usually have diarrhea. Most of such cases may linger for weeks, meanwhile scattering the poison in the discharges. The disease may be recognized in these cases as soon as they are observed to act suspiciously, and there should be no delay in determining at once the nature of the disease. When the animal has been opened the large intestine should be carefully slit up and examined, beginning with the blind or upper end. There will be seen roundish, yellow or blackish spots, having an irregular, depressed, sometimes elevated surface. These spots correspond to dead portions of the mucous membrane, and they are frequently seen from the outside as soon as the animal is opened. Sometimes the membrane has been entirely destroyed. (See also pp. 39–52.)

In order to comprehend fully the reasons for the preventive measures suggested, let us briefly trace the various ways in which hog cholera bacteria may pass from a diseased or dead animal to a healthy one.

Pigs may become directly infected by feeding on the carcasses of such as have died of the disease, or by eating food contaminated with the feces and urine of sick animals, or they may become indirectly infected by feeding upon material in which hog cholera bacteria are accidentally present, and in which they have multiplied. This would include milk, water, and perhaps most vegetables in a boiled condition. It has been pointed out in preceding pages that hog cholera bacteria multiply very abundantly in milk, especially in warm weather, that they remain alive in water for months, and that they multiply upon boiled potato. It has also been shown by an extended series of experiments (p. 80) that they may remain alive in the soil for from one to four months. The sources of infection are thus numerous enough. It has likewise been demonstrated that these disease germs will resist drying. Hence, dried discharges of the sick or the dried body of dead animals are still infectious.

The channel of infection is in most cases the food and drink. This has been frequently demonstrated and emphasized in these pages.

The food, after leaving the stomach, passes in a liquid condition through the small intestine, so that this never seems filled; in fact, its only contents are a coating of semi-liquid matter over the mucous membrane. It passes through the small intestine quite rapidly, but on reaching the large intestine the undigested remains become more consistent, because the liquid is re-absorbed, and are kept here for some time. The bacteria, if not destroyed by the gastric juice, pass quickly through the small intestine, but in the large intestine they begin to multiply and attack the mucous membrane, which they destroy. Thus the feces or discharges of diseased pigs, wherever deposited, scatter larger or smaller quantities of the virus in this way, completing the circle of infection.

In order to prevent the remaining healthy animals in an infected herd from taking the disease the following measures are suggested as of importance, some or all of which may be carried out according to circumstances:

(*a*) *Removal of still healthy animals to inclosed uninfected ground or pens, as far as possible from infected localities.*

(*b*) *Destruction of all diseased animals.*

(*c*) *Careful burial or burning of carcasses.*

(*d*) *Repeated thorough disinfection of the infected premises.*

(*e*) *Great cleanliness, both as to surroundings and as regards the food, to prevent its becoming infected.*

(*a*) The importance of this measure need not be insisted upon, after what has been stated of the various ways in which pigs may be infected. The distance to which they may be removed will, of course, depend on circumstances. They should be kept so far away that there can be no means of communication, either by direct contact, by drainage of the surface of the soil, or by gusts of wind. They should not be kept too closely confined, for if the disease should have attacked one or more and not manifested itself before removal, the infection would become general. Even after this precaution is taken, latent disease among such as are apparently healthy may infect the new grounds and the remaining healthy animals. This danger is increased by the fact that not unfrequently a number of animals become infected from the same source at the same time. Some will show symptoms very speedily; in others the disease will remain latent for a longer time. Under such circumstances it is impossible to properly isolate the well from the sick. Then there is the difficulty of preventing the well animals from carrying the virus on the skin and feet into their new quarters. These drawbacks may be in part overcome by very prompt action when the first signs of disease appear in a herd, before the virus has had an opportunity of being scattered about. The bodies of those to be removed may be fairly well disinfected by pouring over them a 2 per cent. solution of carbolic acid, and forcing them to walk through such a solution.

(*b*) This measure is recommended to prevent the further spread of the virus by the diseased animals. In view of the fact that few recover, that even these few are stunted and of little value, that there is no reliable means of treatment which will eventually cure, destruction of all sick animals is the simplest and most economical procedure in the end.

(c) The disposal of carcasses has already been discussed (p. 128). This very important measure should never be lost sight of.

(d) Among the various disinfectants which can be recommended are the following:

No. 1. Slaked or unslaked lime, used both as a powder and as slaked lime, containing about 5 to 10 per cent. of dry lime (from ½ to 1 pound of lime to a gallon of water).

No. 2. Crude carbolic acid, prepared by adding to the crude carbolic acid obtainable from druggists at about 90 cents a gallon an equal quantity of ordinary sulphuric acid. This mixture is to be carefully added to water in the proportion of 2 ounces to 1 gallon of water, about 1¼ per cent. volume (see p. 91).

No. 3. A 1 per cent. solution (volume) of ordinary sulphuric acid (1¼ ounces of the acid to 1 gallon of water).

No. 4. A 2 per cent. solution of pure carbolic acid. This is prepared by heating the crystals slightly until they melt and adding the resulting liquid to hot water, in the proportion of 1¼ ounces to half a gallon of water. (A pound of carbolic acid, crystallized, retails at 55 cents.)

No. 5. Boiling water.

The careful laboratory experiments with these disinfectants, upon which their practical application is based, are given on p. 87. We shall confine ourselves in this place to a description of their employment.

Disinfectants are substances which, in solutions of a certain strength, are capable of destroying disease germs. Consequently they should be applied wherever the disease germs are supposed to be. In case of hog cholera they are attached to the sides and floorings of pens and to the various utensils used in cleaning them. They are mixed with the earth over which the diseased animals have run, or in the water which they have frequented. In the report of the Department for 1886 the use of mercuric chloride (corrosive sublimate) was recommended, as it is a powerful disinfectant. Since that time other disinfectants have been tested which are equally cheap and easily procurable. The main objection to mercuric chloride is its extremely poisonous character, which makes it undesirable to deal with. This substance has, therefore, in spite of its powerful germicide properties, been thrown out of our list of available disinfectants.

The wood-work of pens, fences, flooring, etc, is best disinfected by using upon it, with a broom, solution No. 2 until thoroughly wet. In preparing this solution it should be stated that the mixing must be done in a glass bottle or jar and the mixture poured slowly into the proper amount of water in a wooden pail. This should be rinsed out after using to prevent the acid from slowly destroying the iron hoops.

Whenever No. 2 is not obtainable No. 3, which seems to be equally efficient, may be used in its place.

Lime is a very efficient disinfectant for hog cholera. Experiments have shown (p. 93) that a solution containing only .02 per cent. will destroy the bacteria. When much organic matter is present, as much as .5 per cent. to 1 per cent. may be necessary. We recommend the proportions given under No. 1, which give from 10 to 20 times the strength required. The resulting liquid is not too thick to be easily manipulated It may be used on wood-work as a whitewash, and it may be spread as a thin layer over the soil which has been infected.

The 2 per cent. solution of pure carbolic acid should be used whenever No. 2 may act injuriously by virtue of the sulphuric acid which it contains.

In general we recommend the use of No. 2 or No. 3 as often as it may seem necessary. It should be dashed upon the infected pens, troughs, tools, and over the infected soil. When there is no objection to lime this may be used on the soil, as it is equally efficient. The discharges should be covered with powdered or slaked lime, and this should be thrown in abundance into pools or wherever water stagnates. In the case of troughs for feed, iron tools, etc., which are liable to injury, the disinfectant should be washed away with boiling water if this is at hand. Boiling water will destroy hog cholera germs by simple contact, and the disinfection will thus be made more complete. Shoes may be disinfected by rubbing them with solution No. 4.

It must be carefully borne in mind that no manure from sick pigs should be removed until it has been treated with disinfectants. The cleaning must be done *after* disinfection and not *before*, to prevent the dissemination of living virus.

The agency of mice and rats in transporting virus will depend upon the promptness and thoroughness with which disinfection and cleanliness are practiced. Mice are more dangerous than rats, in so far as they may take the disease by feeding (p. 72). Flies can only carry such small quantities of the virus that they are not likely to prove dangerous if disinfection and cleaning of feed troughs are attended to.

If these measures are carefully carried out, the disease may either be checked or else mild cases only will appear, owing to the small quantity of poison which the animals are likely to consume with the food.

The epizootic may be terminated by the destruction of most of the herd. This unfortunately is usually the case. What precautions must be taken to prevent subsequent outbreaks?

If only a few animals remain, it is best to slaughter them; they are likely to suffer with the disease in a mild form and continue to infect the premises. If no more animals remain, there should be a final thorough disinfection and subsequent cleaning of the whole exposed territory, including every nook or corner where the disease has existed. This should be done with solution No. 2 or No. 3 as directed, used as freely as possible. After one or two days the ground may be covered with a thin layer, one eighth inch or more, of slaked lime in the strength above given and left undisturbed. If there is no objection to whitewash, this may be applied to infected wood work as an additional safeguard. Even after this thorough treatment, it is best not to place any fresh pigs on the premises for at least four months after the final disinfection. When animals still remain that have been exposed and have not taken the disease, no fresh animals should be introduced for at least six months after the termination of the outbreak. The disinfection must have been equally thorough.

After all this trouble has been taken there is still remaining the danger of a fresh introduction of the disease, and we would therefore again call attention to the rules laid down in the beginning of this chapter. These, after all, must be considered as most important. It is much easier to keep the disease away than to eradicate it after it has been introduced, without great loss of time and money. We would also suggest that in those regions where the danger from hog cholera epizooties is always present, the methods of keeping hogs be simplified in such a way that disinfection may be practiced without too much labor and uncertainty as to the results. It is only necessary to visit a few farms to be convinced of the difficulty that may be met with in endeavoring to eradicate the disease. The hogs are allowed to stray into the most out-of-the-way places when sick instead of being kept in inclosures of definite form and size, which are readily accessible. The poison is thus scattered in such a way as to make disinfection impossible. It is certainly not necessary in raising pigs to allow them to stray into arbors, behind hedges, hide themselves under barns and outhouses. In some farms which we have visited, and which were said to have hog diseases most of the year, there seemed to be no places about the house or garden where pigs did not go. Under such circumstances disinfection is is quite impossible. The pens and other wooden structures, fences, etc., are also apt to be in a very dilapidated condition, so that cleaning is very much complicated. Even under such circumstances the germs will finally perish without disinfection if enough time be given, since they gradually die in the soil and water, as our experiments have shown. A period of from six to nine months after all animals have been removed will be, in general, sufficient to purify the soil of these disease germs. In fact the natural disinfection is very probably accomplished in many cases in from three to four months, but it would not be safe to rely upon this.

TREATMENT OF HOG CHOLERA.

Upon this subject very little should be said, for the reason that diseased animals are a source of poison and a menace to healthy animals, and should be destroyed. Moreover treatment is exceedingly unsatisfactory, as the disease either terminates fatally, whatever remedies are used, or generally makes the animal useless if it should recover. We therefore urgently recommend slaughter of the sick and thorough disinfection as the safest and most economical treatment in the end.

Treatment, however, is resorted to by a large number of owners of swine. The number of specifics, so-called, which are being advertised is legion. We have tried some of the best recommended and found them of no avail. Nor is there any "specific" known in the range of veterinary or human medicine that will cure diphtheritic and ulcerative disease of the large bowels except time, combined with careful dieting, rest, and a few palliatives to relieve pain. It is impossible to carry out this treatment on swine. The success frequently reported with specifics in hog cholera are very probably due to the fact that the treatment is usually adopted in mild forms of disease of a different nature which is mistaken for cholera, or when the acute outbreak is over and the plague has assumed a chronic character. The affected swine linger for a time with very slight symptoms of disease, and this change is credited to the "specific" employed. Remembering that the severest injury is done to the walls of the large intestines in this disease, we regarded it important to determine what medicine would give a prompt and copious evacuation of the bowels in the very beginning of the disease. Various medicines were tried by Dr. Kilborne, at the Experiment Station, among others the following:

(1) *Calomel and jalap.*—February 20, 1888. To No. 463. 30 grains calomel; to No. 467, 23 grains jalap; to No. 468. 30 grains each of calomel and jalap. February 21. Same dose given again to Nos. 463 and 468; no result.

(2) *Calomel.*—March 7. To Nos. 441 and 442, each one dram of dry calomel. March 8. To 441 same dose mixed with castor-oil; to 442 about 1½ drams with castor-oil. No. 441 was freely purged after second dose, continuing for twenty to thirty-six hours. No. 442 was freely purged after sixteen hours, continuing sixteen to twenty hours, when it died. At autopsy were found intense inflammation of stomach, superficial necrosis of the mucosa of large intestine with deep reddening resembling hog cholera. No bacteria found in cultures from spleen. These changes were no doubt due to the calomel.

(3) *Calomel.*—March 8. Nos. 463 and 468 received each one dram colomel in 2 drams castor-oil. No. 468 was purged freely in twenty hours, continuing thirty-six hours. No. 463 was purged in sixteen hours and made ill for several days.

(4) *Epsom salts.*—Nos. 403 and 405 (weight 50 pounds) received each 1 ounce. Bowels slightly relaxed for one or two passages. Nos. 339 and 377 (weight 50 pounds) received each 2 ounces in water. No. 339 was purged and made slightly ill. No effect on 377.

(5) *Barbadoes aloes.*—Nos. 402 and 404 (weight 50 pounds) received one-half dram each; no effect. Nos. 372 and 380 (weight 65 pounds) received each 2 drams, mixed in molasses; no effect. The same animals, five days later, received each 4 drams with molasses; no effect, except discoloration of feces.

(6) *Castor-oil and turpentine.*—Nos. 387 and 388 (weight 50 to 60 pounds) received each 1⅓ ounces castor-oil and one-sixth ounce turpentine; no effect. No 387 received, five days later, 2⅓ ounces oil and one-sixth ounce turpentine. No. 388 received, five days later, 2⅓ ounces oil and one-third ounce turpentine; no effect.

(7) *Linseed-oil and turpentine.*—Nos. 383 and 399 (weight 50 to 60 pounds) received each 2⅓ ounces oil and one-sixth ounce turpentine; no effect. No. 383 received, five days later, 4 ounces oil and one-sixth ounce turpentine. No. 399 received five days later, 4 ounces oil and one-third ounce turpentine. Both were made sick for a day or two. No catharsis.

These trials show how difficult it is to cause movement of the large bowels in swine, and they also suggest that this very sluggishness may make them susceptible to inflammations and ulcerations such as we find in hog cholera and frequently in swine plague.

It was our intention to obtain a cathartic which would freely purge without causing any inflammation or irritation of the large intestine. Of those tried, calomel is the only available one. This must be carefully given, as it may produce the very inflammations which it is designed to check, and destroy life, as was actually done in experiment 2.

Concerning calomel Ellenberger* says:

Calomel (in combination with castor-oil) is especially serviceable with swine; with larger animals when the contents of the intestinal canal are to be disinfected, and in inflammatory fevers. It should be given to ruminants with the greatest caution.

It was our purpose to try calomel after having made these trials upon healthy animals, when the disease died out at the Experiment Station, and further investigations had to be postponed.

If the large intestine has been promptly evacuated the next important step is to give only that food which leaves but little irritating waste to pass into the large bowel, such as milk and gruels. In short, it is best to use only boiled or scalded food so as to help the process of digestion as much as possible. It may be necessary to repeat the dose of calomel after a few days. As to this mode of treatment our experience is not sufficient to warrant any positive statements, and it is simply suggested to those who wish to run the risk of treating this disease.

* *Lehrbuch d. allgemeinen Therapie d. Haussängethiere*, 1885, p. 676.

There is another line of preventive and curative treatment which may prove valuable in the future, namely, the feeding of substances with the daily food which, while not injurious to the animal itself, may keep in check the multiplication of the virus in the intestine by an antiseptic action. It is very important, however, to bear in mind that a large number of those medicines which act as disinfectants and antiseptics are likewise injurious or even poisonous to the animal itself. A too abundant feeding of such material, while it may reduce the mortality and lessen the severity of the disease in the sick, is liable to cause injury to liver, kidneys, and other vital organs, whereby the nutrition of the animal may be permanently injured. Such medicines, when carelessly given to healthy animals as preventives, may irritate the large bowel sufficiently to reduce its vitality and power of resistance when the disease actually appears. The proper medicine to feed must therefore be selected with care, and we trust that experiments to this effect may be carried on at the Experiment Station at an early date.

There is still another line of treatment which demands attention, namely, the introduction of a sufficient amount of some disinfectant into the body to be absorbed and thus to make the whole body oppose the multiplication of bacteria. Koch tried this method by injecting mercuric chloride into guinea-pigs and afterwards inoculating them with anthrax bacilli. The animals all took the disease and died.

At the laboratory of the Bureau mercuric iodide, a still more powerful disinfectant, was tried upon rabbits.

A solution was prepared containing .001 gram mercuric iodide and .002 gram potassic iodide in a cubic centimeter. Of this .5 cubic centimeter was injected beneath the skin of the back of four rabbits (Nos. 1, 2, 3, 4) for eight successive days. On the third day Nos. 1, 2, 3, and a fresh rabbit (check), No. 5, received hypodermically into the inner surface of the thigh ¼ cubic centimeter of liquid containing one-millionth cubic centimeter of a beef infusion culture of hog cholera bacteria about one day old.

All the inoculated rabbits died, the dates being given in the appended table. Rabbit No. 4, which had received the iodide only, to observe any poisonous effect, remained well. The lesions were those of hog cholera, and the specific bacteria were present in the spleen. The total amount of the iodide given was .004 gram, about $\frac{1}{16}$ grain.

No.[*]	.5 cubic centimeter of $\frac{1}{10}$ per cent. mercuric iodide daily.	Inoculated with one-millionth cubic centimeter culture hog cholera bacteria.	Remarks.
1	June 22 to 29, inclusive	June 24	Died June 30.
2	....do	....do	Died July 3.
3	....do	...do	Died July 6
4	...do		
5	... do	June 24	Died July 3.

* Weighing each about 2 pounds.

At the same time healthy pigs were fed with the same substance in minute doses, to observe any toxic effect that might appear. These experiments were likewise interrupted in their application by the disappearance of the disease at the beginning of the year. While we therefore recommend in general the use of a purgative, such as calomel, in the beginning of the disease, and careful feeding subsequently, we have as yet no actual experimental evidence that such treatment will be of any avail, owing to the frequent interruptions of the work.

EXPERIMENTS ON THE PREVENTION OF HOG CHOLERA BY INOCULATION.

During the past six or seven years the attention of the world has been centered upon the brilliant experiments of Pasteur in the line of preventive inoculation for animal diseases. Many others have since then followed his footsteps with variable success. Among those diseases for which preventive inoculation is more or less in vogue in some European countries, especially in France, are anthrax among cattle and sheep, *charbon symptomatique*, affecting the same animals and perhaps identical with the disease known as black quarter or black leg in this country. This is not the place for discussing the absolute value of vaccination in these diseases, nor to point out the reasons why they are not regarded of much value by scientists of Germany and other States. Suffice it to say that the Bureau has devoted a large amount of time in testing all the available methods of attenuating hog cholera virus for vaccinal use, to determine whether any vaccine could be placed at the disposal of the public that might prevent the enormous losses entailed each year by this dreaded germ disease. In the following pages most of the work done (chiefly during the winter and spring of 1886) is briefly reproduced.

Inoculations of small doses of strong virus *in the form of liquid cultures* was first resorted to, because it had been observed that pigs rarely take the disease when culture liquid in such quantities is injected directly under the skin. We reasoned that if injections of small doses do not cause the disease, why may they not prove effective in preventing natural infection? The cultures were not therefore attenuated, as Pasteur has done, with anthrax, because the step was deemed unnecessary.

A lot of animals were at first inoculated twice with very small quantities, the period between the two inoculations being about two weeks. This time was sufficient to reveal any disease which might have been induced by the inoculations. Two weeks after the second inoculation the animal was infected either by allowing it to feed upon the internal organs of pigs which had died of the disease or by exposing it to the sick and dying in an infected pen. It was soon found that the inoculations were by no means protective, in whatever way the virus entered the system, for subsequent feeding usually produced cases of the most acute character and with the most severe and extensive lesions. The doses of inoculated cultures were gradually increased in quantity without yielding any better results. Of a large number of animals subjected to inoculation only five took the disease unmistakably as a consequence of the operation. The experiments, including tables and *post-mortem* notes, are given *in extenso* as they were made.

In reading them over it will be noticed that the virus was cultivated chiefly in liquid media, and the solid media, more particularly nutritive gelatine, were only employed to test the purity of the cultures. Whenever these cultures were used for inoculations they were previously tested on gelatine plates by drawing a platinum wire, dipped into the culture, through the gelatine layer two or three times before the gelatine had become solid. Among the hundreds of cultures thus tested in the space of several months not one was found impure. Series of cultures extending up to the tenth generation were usually carried on by inoculating fresh tubes each day. The last culture tested as described above gave precisely the same colonies as the first in all the series thus far prepared. The culture-tube, described in the First Annual Report of the Bureau, was used almost exclusively for these cultures in liquid media. The advantages and accessibility of cultures in liquids for purposes of inoculation, the readiness and ease with which quantities or doses may be determined, finally, certain characteristics of growth in liquids, place this method on a level with, if not above, that of solid cultures for experimental purposes. For diagnostic purposes, solid media are to-day a *sine qua non* of bacteriological work.

Experiment 1.—Pigs Nos. 152, 167, 168, and 175 were inoculated with pure cultures in beef-infusion peptone as follows: On January 23, one drop of the seventh culture, derived from the spleen of pig No. 114; on February 8, with one eighth cubic centimeter from a culture derived from a guinea pig. Both cultures were diluted in sterile normal salt solution in such a way that 1 cubic centimeter of fluid was injected each time. The inner aspect of the thigh near Poupart's ligament was chosen, and the liquid was introduced beneath the skin into the subcutaneous tissue with a hypodermic syringe. There was no perceptible swelling at the site of either inoculation, excepting in No. 175, in which there were two tumors, each the size of a walnut, at the seat of the first inoculation. In order to test the extent of the immunity which these inoculations might have conferred, feeding the viscera of pigs which had succumbed to hog cholera was resorted to, the animals being transferred to the large infected pen for this purpose. Nos. 168 and 175 were fed in this way March 5, and two animals not inoculated (Nos. 158 and 189) were fed with them. All four died; the two vaccinated animals in about twenty days, the others in about fifteen days after feeding. March 13, Nos. 152 and 167 were fed with two check animals, Nos. 176 and 190. These four also died of hog cholera; the two vaccinated ones averaging twenty days, the others eleven days after feeding. The inoculation may be said to have simply retarded death from five to nine days. A table giving a summary of these facts is appended, together with a brief description of the *post mortem* appearances :

No.	January 23.	February 8.	Fed with bog-cholera viscera.	Died.	Number of days after feeding.
	Drops.	*c. c.*			
152	1	½	Mar. 13	Apr. 3......	21
167	1	½	Mar. 13......	Apr. 1......	19
168	1	½	Mar. 5......	Mar. 28.....	23
175	1	½	Mar. 5......	Mar. 22.....	17
158	....	....	Mar. 5......	Mar. 21.....	16
189	....	....	Mar. 5......	Mar. 19....	14
176	....	....	Mar. 13	Mar. 23....	10
190	....	....	Mar. 13....	Mar. 25. ...	12

Autopsy notes.—No. 152. Skin of limbs and abdomen dotted with purple spots; on abdomen, general reddening. Points of extravasation and ecchymosed spots throughout the subcutaneous connective and fatty tissue and on gastro-splenic omentum. Superficial inguinal glands greatly enlarged and congested. Spleen enlarged, filled with blood, and very soft. Petecchiæ on epicardium. Numerous lobules of the lungs collapsed. Glomeruli of kidneys appear as deep red petecchiæ. In cæcum and upper portion of colon extensive and deep ulcers. A few in the ileum near the valve. The mucosa of the stomach, small and large intestine, thickly covered with dark red points or petecchiæ.

No. 167. Dying, and hence killed by a blow on the head. Spleen swollen, friable; epicardium dotted with points and spots of extravasation. In lungs a few collapsed lobules. Lymphatic glands generally very deeply congested, similarly the mucous membrane of fundus of stomach and the kidneys. Large ulcers in cæcum and upper portion of colon.

No. 168. Subcutaneous and sub-peritoneal tissue contains numerous ecchymoses from one-eighth to three-fourths inch in diameter. Spleen enlarged, gorged with blood, friable. Petecchiæ on epicardium. Lungs not collapsed; its parenchyma contains numerous deeply-congested areas from one-eighth to one-half inch in diameter. Kidneys enlarged, with extravasations on surface and in parenchyma. Cortex of lymphatics in general deeply congested. Extensive, almost continuous, ulceration of cæcum, and upper portion of colon, in part blackish, the remainder of the large intestine being the seat of severe inflammation and extravasation. Mucous membrane of stomach similarly involved.

No. 175. Subcutaneous tissue dotted with pale red spots. Tumor at the place of the first inoculation firm throughout, pale yellowish. Superficial inguinal glands, as well of those of thorax and abdomen, with purplish cortex. Spleen tissue still firm, dotted with numerous bright red points, but slightly enlarged. Beneath the entire epicardium and endocardium many extravasations. Cæcum and upper portion of colon extensively ulcerated. Serous surface of large intestine dotted with extravasations.

No. 176. Slight reddening of skin and subcutaneous fatty tissue. Cortex of lymphatic glands in general deeply congested. Spleen much enlarged and surface dotted with numerous bright red elevated points. A few petecchiæ on endocardium and epicardium. Lungs deeply congested throughout; kidneys likewise inflamed. Stomach slightly reddened at fundus; small intestine also slightly congested. Serosa of large intestine dotted with extravasations. The mucosa of cæcum and small portion of colon one mass of necrosed tissue. Walls thickened.

No. 189. Extensive and deep reddening of skin of abdomen, throat, and limbs. Subcutaneous tissue only slightly reddened; spleen enlarged, gorged with blood, friable. Besides the general congestion of the lungs there are small darker areas, representing hemorrhagic lobules. Bronchial glands and those along lesser curvature of stomach swollen and gorged with blood; the other lymphatics only moderately congested. Besides a small number of ulcers throughout the large intestine, the mucous membrane is deeply congested and dotted with occasional hemorrhagic points. Kidneys extensively inflamed; on section the cortex shows extravasations.

No. 190. Considerable reddening of the skin of abdomen and ventral aspect of limbs; very slight in subcutaneous tissue. Spleen greatly enlarged, dark purple; blood flows freely on cutting into it; very soft. Lungs contain regions of congestion and hepatizations, possibly due to the presence of a few lung worms. Lymphatic glands near stomach, the bronchial and superficial inguinal glands, deeply congested. Mucous membrane of stomach extensively congested; a large patch of extravasation in fundus; large intestine severely inflamed, with occasional extravasations; no ulcerations.

The diagnosis of hog cholera was confirmed in every case by finding the specific bacillus in cover-glass preparations of splenic tissue and obtaining therefrom pure cultures in liquid media and in gelatine.

Experiment 2.—In conjunction with the first series of inoculations, two pigs (Nos. 149 and 161) were inoculated at the same time, as follows: January 23, with 1 cubic centimeter of the seventh culture in beef-infusion peptone. No reaction at the place of inoculation in No. 149; a tumor as large as a marble in No. 161. On February 8 both received a second injection of 1 cubic centimeter. Two swellings as large as a chestnut at the place of the second inoculation in No. 149; in No. 161 also a considerable thickening was present. No. 149 was fed March 5 with four of the preceding series; No. 161 on March 13 with the remainder of the preceding series and some to be subsequently spoken of. Both died of hog cholera. The accompanying table and brief autopsy notes explain themselves:

No.	Inoculation.		Fed.	Died.	Days after feeding.
	Jan. 23.	Feb. 3.			
	c. c.	c. c.			
149	1	1	Mar. 5	Mar. 24	19
161	1	1	Mar. 13	Apr. 14	32

No. 149. Slight reddening of the skin and subcutaneous connective tissue; the tumors produced by inoculation firm, pale yellowish, only one showing softening within; spleen considerably enlarged and full of blood; ascarides in gall bladder, which is ulcerated; mucous membrane along fundus of stomach intensely congested; the mucous membrane of cæcum and upper portion of colon one mass of ulcers; in the remainder of colon they are isolated; kidneys congested.

No. 161. Great emaciation; spleen enlarged and gorged with blood, very soft; all excepting the posterior region of each lung hepatized and the bronchi filled with a thick creamy mass, which consists almost entirely of pus corpuscles; lymphatics but slightly congested; adhesions between adjacent coil of large intestine and bladder; cæcum and colon studded with large deep ulcers; valve greatly enlarged; intense congestion of mucous membrane of fundus of stomach.

Cover-glass preparations from the spleen of both contain the characteristic bacilli. Gelatine and liquid cultures from the same organ were pure.

The comparatively large dose of strong virus used for vaccination was not capable of protecting these animals from the disease communicated by feeding. There was no suspicion of disease caused by the vaccination when they were fed, and the time intervening between the two injections was sufficient for any development of disease from the injected virus.

Experiment 3.—Pigs Nos. 151, 169, 170, and 178 were inoculated as in the preceding experiments on February 8 and 23 with $\frac{1}{8}$ cubic centimeter of a beef infusion peptone culture derived from a guinea pig and the seventh culture from the spleen of a pig in the same medium. The dose was diluted in salt solution so as to make 1 cubic centimeter of liquid. In No. 151 the second inoculation produced a tumor about 1 inch long and one third inch thick. The first was scarcely noticeable. In No. 169 the first inoculation resulted in a bean like nodule; the second produced several of the same size. In No. 170 neither inoculation showed more than a very slight swelling. In No. 178 both inoculations produced rather extensive swellings.

On being fed with the viscera of pigs known to have died of the disease all took the disease and died; two on March 13 and the remaining two on March 19, *i. e.*, one in thirteen, one in eighteen, and two in twenty-two days after feeding. A table summarizing these facts and brief *post-mortem* notes are appended:

No.	Inoculation.		Fed.	Died.	Days after feeding.
	Feb. 8.	Feb. 23.			
	c. c.	*c. c.*			
151	$\frac{1}{8}$	$\frac{1}{8}$	Mar. 13	Mar. 26	13
169	$\frac{1}{8}$	$\frac{1}{8}$	Mar. 19	Apr. 10	22
170	$\frac{1}{8}$	$\frac{1}{8}$	Mar. 13	Apr. 4	22
178	$\frac{1}{8}$	$\frac{1}{8}$	Mar. 19	Apr. 6	18

Autopsy notes.—No. 151. Purplish spots on skin of abdomen and paler ones in subcutaneous tissue. Inoculation tumor cuts like cheese; yellowish-white. Extravasations under endocardium and pericardium; left lung mottled from hemorrhagic areas; cortex of lymphatic glands infiltrated with blood; those of meso colon and lesser curvature of stomach dark purple throughout; kidneys pale; hemorrhage into pelvis of left kidney; extravasations into mucosa of stomach; moderate number of ulcers in cæcum and colon; large quantity of blood in the lower 6 or 8 feet of ileum and in the large intestine, clotted in the former tube, where the mucous membrane is deeply congested.

No. 169. Small tumor on the left side, the place of the second inoculation; spleen enlarged and congested, with large hemorrhagic infarcts; considerable effusion in the large serous cavities. Besides the general congestion of lungs, there are scattered throughout its parenchyma hemorrhagic foci. Hemorrhagic inflammation of kidneys manifested by bright red glomeruli throughout its cortex; lymphatics in general deeply reddened; numerous petecchiæ in stomach, small and large intestine. In cæcum and colon, large, deep ulcers.

No. 170. Redness of skin in abdomen; nothing at places of inoculation; spleen enlarged, friable, full of blood; abdomen, thorax, and pericardial cavity contain much yellow serum; congestion of the lungs with darker hemorrhagic foci throughout; anterior lobes collapsed; kidneys enlarged, with a few extravasations on surface and in parenchyma; mucous membrane of stomach and intestines covered with many hemorrhagic points and spots. In large intestine, including rectum, numerous old ulcers, some 1 inch across. Lymphatics in general hemorrhagic.

No. 178. Died quite unexpectedly. At the place of first inoculation two firm whitish masses; spleen enlarged, friable; its substance contains hemorrhagic infarcts; extravasations beneath both serous surfaces of the heart; congestion of lungs, with numerous darker hemorrhagic foci; lymphatic glands of abdominal cavity very dark and gorged with blood; extensive ulceration about the ileo-cæcal valve, in the cæcum, and colon; in the lower portion of colon and in the rectum numerous small extravasations. Hemorrhage into pelvis of both kidneys.

The *post-mortem* determination of a severe type of hog cholera in these four cases was confirmed by finding in the spleen of each animal, by means of cover-glass preparations, numerous specific bacteria of this disease. Cultures in liquid media made from every spleen were found pure when examined microscopically as well as on gelatine plates. This experiment likewise proved the inefficiency of small quantities of non-attenuated virus introduced beneath the skin in preventing an invasion of the micro-organism from the alimentary canal.

Experiment 4.—A third lot of four pigs (Nos. 117, 171, 172, and 174), between three and five months old, were inoculated as before with .2 cubic centimeter each from the second beef infusion peptone culture derived from a pig's spleen. On March 1 they were inoculated with .2 cubic centimeter from the second culture derived from a pig's spleen. In No. 117 there was a slight swelling after the first, and one as large as a chestnut after the second inoculation. In No. 171 a mass 1½ to 2 inches long and three-fourths inch in diameter was found at site of the first inoculation. There was but a small nodule at the place of the second inoculation. In No. 172 two lumps, like small marbles, formed after the first inoculation; after the second only a small nodule formed. In No. 174 the reaction after the second inoculation was manifested by an irregular tumor about 2 inches long and one third of an inch in diameter, the reaction at the place of the first inoculation being less marked.

Of these four, two (Nos. 117 and 172) were fed with the viscera of pigs dead from hog cholera, together with two control animals (Nos.

192 and 193), on March 19. The rest (Nos. 171 and 174) were simply placed in the large infected pen March 22, with those that had been fed with infectious matter. Below the result is given in a tabulated form. It shows that all the animals succumbed to the disease, those simply exposed by contact with the sick as well as those fed. Of the inoculated animals, those fed died in twenty-one and eighteen days after feeding; those exposed, in twenty-two and twenty-five days, respectively. Those not inoculated died twelve and nine days, respectively, after feeding. Here, likewise, we notice the prolongation of life in the inoculated pigs.

No.	Feb. 13.	Mar. 1.	Date of feeding and exposure.	Died.	Days after exposure and feeding.
	c. c.	c. c.			
117	.2	.2	Fed Mar. 19........	Apr. 9	21
171	.2	.2	Exposed Mar. 22..	Apr. 13	22
172	.2	.2	Fed Mar. 19.......	Apr. 6	18
174	.2	.2	Exposed Mar. 22..	Apr. 16	25
192*	....	....	Fed Mar. 19.......	Mar. 31	12
193*	....	....	......do	Mar. 28	9

* Checks.

The lesions found at the autopsies of these pigs are briefly as follows:

No. 117. Extensive reddening of the skin of abdomen; great enlargement of spleen, which is gorged with blood, very soft; petecchial discolorations on surface of lungs and on section; large intestine studded with broad, deep ulcers as far as the rectum; a few in ileum.

No. 171. Skin over ventral aspect of body deeply reddened; hemorrhagic spots under peritoneal covering of diaphragm and large intestine and under capsule of kidneys; lungs congested, containing numerous dark hemorrhagic lobules; part of anterior lobes collapsed. The spleen very large, dark colored; nodes slightly raised above surface, shown on section to be hemorrhagic infarcts; lymphatic glands generally highly congested; petecchial spots on surface and in cortex of kidneys; hemorrhagic foci throughout mucosa of stomach and intestines. About four large ulcers in cæcum and colon.

No. 172. Reddening of skin of ventral aspect of body and of subcutaneous tissue generally; firm, pale yellow, cheesy masses, surrounded by a thin membrane at place of inoculation; engorgement of spleen and lymphatic glands; extravasations in parenchyma of kidneys. In cæcum and colon numerous deep ulcers, some coalesced. Mucosa of stomach generally congested, and that of intestines thickly dotted with petecchiæ.

No. 174. Deep reddening of skin of abdomen; encysted cheesy mass at site of first inoculation; great enlargement of spleen; prominent red points on surface; effusion into abdominal cavity; anterior lobes of lungs collapsed, remainder normal; lymphatics highly congested; three large ulcers in cæcum; valve thickened and ulcerated; petecchiæ numerous throughout mucosa of stomach and intestines.

No. 192. Control animal; reddening of skin of ventral aspect of body and of subcutaneous tissue; spleen swollen; full of blood; friable; atelectasis of the small anterior lobe of each lung; ulcers on the mucous surface of gall bladder; cortex of lymphatic glands infiltrated with blood; mucosa of large intestines congested; numerous ulcers in caecum and upper colon.

No. 193. Subcutaneous connective tissue considerably reddened; spleen but slightly enlarged; not much softened; mucous membrane of stomach, of large and small intestines, deeply congested; contents of large intestine fluid, chocolate colored.

In cover-glass preparations from the spleen pulp of these animals, numerous bacteria of hog cholera were found in each preparation. Both gelatine and liquid cultures from every spleen proved to be pure cultures of the bacillus of hog cholera.

The diagnosis made on *post mortem* was thus confirmed by microscopic examination and culture.

Experiment 5.—To determine the effect of a single inoculation, on February 13 two pigs (Nos. 115 and 160) received subcutaneously each 1 cubic centimeter of the second beef infusion peptone culture obtained from the spleen of a pig. In No. 115 a tumor as large as a marble was found at the seat of inoculation March 9. In No. 160 the tumor was elongated, about 2 inches long and three-eighths of an inch thick. No. 115 was fed with viscera taken from cases of hog cholera March 19. No. 160 was simply exposed to the disease by being transferred to the large infected pen. No. 115 died April 8. No. 160 recovered and was well May 6. The detailed account of this experiment is appended:

No.	Feb. 13.	Date of feeding and exposure.	Effect.	Days after feeding.
	c. c.			
115	1	Fed Mar. 19	Died Apr. 8 .	29
160	1	Exposed Mar. 22....	Recovered ..	

Post-mortem notes.—No. 115. Firm, pale yellow tumor at seat of inoculation, encysted; center undergoing softening. Spleen tumefied, very dark and friable. A few extravasations beneath serous coverings of heart. In cortex of kidneys numerous hemorrhagic points; cystic degeneration of right kidney; advanced ulceration of caecum and colon; scattered petecchiæ in mucosa of stomach and small intestine.

No. 160. Was very low for a time, beginning with April 1. It was barely able to stand and its appetite was poor. It rapidly recovered, however, and was gaining flesh in May. Whether the animal was suffering from hog cholera or from the *Sclerostoma pinguicola* (kidney worm), with which some of this lot were found affected, can not be said.

In order to determine whether a single injection of a comparatively large quantity of culture liquid, while not inducing the disease, would protect against the disease itself, the following experiment was performed:

Experiment 6.—Four pigs (Nos. 202, 204, 205, and 212) were inoculated April 2 with 1½ cubic centimeters of a seventh culture in beef infusion with 1 per cent. peptone one day old. Four additional pigs (Nos. 206, 207, 208, and 209) received but one cubic centimeter of the same culture. The remaining four of the same lot (Nos. 203, 210, 211, and 213) were reserved as checks upon the experiment. Of these Nos. 203 and 213 had a temperature of 106° F., and hence were suspected of disease. This suspicion was soon confirmed after they had been placed in a pen alone. Both had a severe diarrhea, one dying April 11, the other April 13. The lesions were confined to the mucous membrane of the large intestine, which was dotted with numerous elevated lemon-yellow tough masses a few lines across, simulating ulcers. On close examination, however, this impression was dispelled. These tough masses were easily removed *in toto* from the mucosa, which presented a slight depression without any loss of substance. They were evidently exudates from the mucosa (croupous?). There were no bacteria in the blood or in a bit of spleen dropped into a culture tube. No development took place in either tube.

Of those inoculated with 1½ cubic centimeters two died from the immediate effects of inoculation. No. 204 died in eleven days and No. 212 in seven days. In No. 204 a tough tumor had formed at the point of inoculation on each side. The mucous membrane of the large intestine was completely necrosed and the spleen enlarged. In No. 212 local swelling was present on one side. The stomach and large intestine were deeply congested, with points of commencing ulceration in the latter. In both animals the bacillus of hog cholera was present in cover-glass preparations of the spleen. Nos. 202 and 205 seemed to remain unaffected by the inoculation. One month and a half later both were exposed to the disease in the large infected pen. A month later they were removed with others to a clean pen, after having apparently resisted infection. No. 202 was gradually wasting away, and died July 24, more than two months after exposure. In the large intestine were cicatrices of healed ulcers and such as were healing. The severest lesions were in the lungs. Both were adherent by means of bands to the costal pleura, and were extensively hepatized. No. 205 was alive and well August 15.

Of the second lot, which had received 1 cubic centimeter of the same culture the results were nearly the same. Two succumbed to the inoculation, one died of infection, and a fourth survived. No. 208 died fifteen days after inoculation. Besides the inoculation swellings, enlarged and congested spleen, the mucous membrane of the large intestine was covered with extensive deep ulcers, and the walls much thickened and softened. The corresponding lymphatics in the meso colon deep purple. No. 209 died in six days after inoculation. There was general congestion and extravasation of blood in the internal organs, involving the entire mucous membrane of the alimentary tract, especially the large intestine, the lymphatics and serous membranes, the spleen, and kidneys. Ulceration had not yet begun. In both animals the spleen was crowded with bacteria and furnished pure cultures of the specific germ.

Nos. 206 and 207 were not affected by the inoculation. They were exposed with the preceding lot, as indicated in the table. No. 207, after apparently resisting infection in the infected pen for a month, died July 18, after having been in a clean pen since June 21. The extensive necrosis of the mucous membrane of the cæcum and upper portion of colon, with the absence of any acute inflammation elsewhere, gave evidence of a chronic case of hog cholera. No. 206, though still alive, is emaciated.

The two remaining check pigs, which were exposed with the preceding animals in the same infected pen, both died of hog cholera; No. 211 found dead June 21. The most marked changes were a small number of ulcers on a pale mucous membrane scattered over the cæcum and colon. No. 210 lived a month longer than its mate. The existence of hog cholera was demonstrated by a general necrosis of the mucous membrane of the cæcum and an extensive pigmentation in the remainder of the large intestine. The lungs were adherent in places and much congested.

When we gather together the facts presented by this experiment we shall find a certain number of interesting deductions springing therefrom. In the first place, we note the peculiarity of the intestinal lesions of the two animals which died from some unknown cause, presumably not hog cholera. We next point to an additional demonstration of the specific nature of the bacillus of hog cholera, for out of eight inoculated four died, and the age of the lesions corresponded well with the length of time elapsing between inoculation and death.

Those animals which resisted the inoculation were in part protected, as two among four were still alive on August 17, and the remaining two died, probably from effects of the ulceration, months after exposure.

No.	Inoculated April 2.	Died from inoculation.	Exposure in infected pens.	Removed from infected pens.	Remarks.
*202	1½ c. c. culture liquid.	....	May 18	June 21	Died July 4.
204	.....do ....;..	April 13	....	....	
205	.....do	....	...do ..	June 21	Alive August 17.
212	......do	April 9	....	....	
206	1 c. c. culture liquid .	....	May 18	June 21	Alive August 11, but unthrifty.
207	... do	....	....	....	Died July 18.
208	.....do	April 17	....	....	
209	.....do	April 8	....	....	
†203	....	....	....	....	Died April 11, from some unknown disease.
†210	....	....	May 18	June 21	Died July 21, of hog cholera.
†211	....	....	...do ...	....	Died June 21, of hog cholera.
†213	....	....	....	....	Died April 13, from same disease as No. 203.

* These animals were one and a half months old at date of inoculation.   † Checks.

Experiment 7.—Having determined that even large doses of liquid cultures of the bacillus of hog cholera can be borne without producing the disease in most cases, it was thought advisable to make two inoculations of strong virus, a first one with a small quantity and a second with a large quantity.

First inoculation, April 21: Nos. 214, 227, 223, and 222 received ¼ cubic centimeter of a third culture in beef infusion containing 1 per cent. each of peptone and glucose. The liquid was diluted with sterile salt solution, so as to make ½ cubic centimeter. It was injected, one-half beneath the skin of each thigh. After waiting two weeks, in order to determine whether the inoculation had not produced disease, a second injection was practiced May 6, the thirteenth and fourteenth cultures of the same series being used for this purpose. The animals received 1, 1½, 2, and 2½ cubic centimeters of the culture liquid, respectively. No untoward results following the injection of these large doses, they were transferred to the large infected pen May 25.

A second lot (Nos. 226, 228, 215, and 229) were treated in precisely the same way and at the same time, excepting in receiving ½ cubic centimeter for the first dose instead of ¼ cubic centimeter.

No.	First inoculation, Apr. 21.	Second inoculation, May 6.	Exposure in infected pen.	Time of death.	Days after first exposure.
	c. c.	c. c.			
214	¼	1	May 25	July 1	37
227	¼	1½	..do....	June 27	33
223	¼	2	...do....	July 2	38
232	¼	2½	...do....	July 1	37
226	½	1	...do....	July 3	39
228	½	1½	...do....	July 13	49
215	½	2	...do....	July 10	46
229	½	2½	...do....	June 27	33
*224	....	....	...do....	Aug. 4	71
*225	....	....	...do...	June 27	33

* Checks.

No. 214, being in a dying condition July 1, was killed. In the cæcum and colon were found very large, deep, blackish ulcers upon a pale mucosa. The case was evidently one of chronic hog cholera. A pure liquid culture of the hog cholera bacillus was obtained from the spleen.

No. 227 died June 27. The lymphatic glands were deeply congested; the mucosa of large intestine was generally pigmented and covered with large blackish ulcers. Small yellowish ulcers were also found in the ileum. The points of injection were occupied by encysted, partly liquefied masses.

No. 223 was found dead July 2. At the points of injection encysted masses were found, the contents of one of which were discharging through an opening in the skin. The mucosa of the entire large intestine deeply congested. Scattered ulcers of varying age and size in the cæcum and colon. Bacilli in spleen.

No. 232, after a period of unthriftiness, was found dead July 1. The autopsy revealed a chronic broncho-pneumonia, with pleuritic adhesions of right lung. The mucous membrane of the cæcum and colon, besides being studded with a large number of shallow ulcers, was deeply and uniformly congested, the congestion involving also the lower portion of the ileum. On both thighs an encysted semi-liquid mass indicated the seat of the inoculation.

Of the second lot, which had received ½ cubic centimeter of the first inoculation, all succumbed to the infection.

No. 226 died July 3. The characteristic lesion was extensive ulceration, together with deep congestion of the mucosa of large intestine. Encysted masses at the points of inoculation. A considerable number of hog cholera bacilli in the spleen.

No. 228 died July 13. In this animal the mucosa of cæcum and colon presented a continuous mass of necrosed blackish tissue, the ileo-cæcal valve being enlarged to twice the normal size. A few scattered yellowish ulcers in the lower portion of the colon.

No. 215 died July 10, probably affected in the same way, though no post-mortem examination was made.

No. 229 died July 27. In this case the lymphatic glands were in general deeply congested; ecchymosis beneath the serous membranes. Pigmentation of the mucous membrane of the stomach, duodenum, ileum, and large intestine from former extravasations. Several large ulcers on the valve and some others in colon. Ulcers in the cardiac portion of the stomach. Encysted masses at the point of inoculation.

Nos. 224 and 225 were penned with the above eight animals as checks. No. 225, after being sick for a few days, was found dead June 27. The mucosa of the cæcum and upper half of the colon was extensively pigmented and ulcerated, the lower half deeply congested. The ileum ulcerated for 5 or 6 feet from the valve. Many of the ulcers were so deep as to have produced inflammation of the serous membrane and thickening of the intestinal walls. The other check (No. 224) lived over two months after exposure, being unthrifty during this period. On post-mortem examination the mucosa of large intestine was considerably pigmented, and scars of healed ulcers were present. A large suppurating wound of the lower jaw, involving the bone, may have contributed towards the fatal issue.

Experiment 8.—These inoculations having failed to produce immunity from natural infection, another experiment was tried by augmenting the dose of strong virus used for the second inoculation. Thus Nos. 239, 242, 244, and 245 received each ½ cubic centimeter for the first inoculation May 27, No. 243 being retained in the same pen as a check. Of these No. 239 died of hog cholera as the result of the inoculation. The remaining three, received two weeks later, on June 10, 2 cubic centimeters each of strong virus. The cultures were prepared in beef infusion with 1 per cent. peptone. They were usually the third or fourth culture, not more than one day old. A second lot (Nos. 240, 254, 255, and 256) were inoculated at the same time and in the same way, with this exception, that the second dose was increased to 3 cubic centimeters. On June 24 all were placed in the large infected pen.

No. 239 died June 2, within six days after receiving ½ cubic centimeter of the culture, and as a result of the inoculation. The lesions were those of a very acute case, engorged spleen and lymphatics, intense congestion of the mucosa of the large intestine and of the intestinal tract in general. The lungs were likewise engorged and dotted with extravasations. This animal was eating and apparently well on the morning of death. The spleen was crowded with bacilli of hog cholera, and pure cultures were obtained from it.

No. 242 died July 17. The characteristic lesions of hog cholera were found in it; extensive ulceration of the cæcum and colon; engorgement of spleen and lymphatic glands with blood. Encysted masses at the point of inoculation. No. 224 succumbed July 9 with practically the same lesions, besides the presence of a considerable quantity of serum in the abdominal cavity.

The check to this lot died July 13. The depth of the ulcerations in the cæcum and colon had implicated the serous covering, so that adhesions had formed between the cæcum and abdominal walls. Punctiform ecchymosis on serosa of ileum; the mucosa not affected. The mucosa of cæcum was found completely ulcerated, the necrosis stopping abruptly at the edge of the valve; in the colon the necrosis resolved itself into large isolated ulcers.

Of the second lot, No. 240 died July 10. At the place of inoculation a firm pale yellowish mass, about 1 inch long, was found. The lower portion of ileum, the cæcum, the upper portion of colon, contained ulcers of different sizes. The duodenum was occluded by a clot of blood. No. 254 died the same day, with lesions of a similar character. No. 255 died July 20. The spleen in this case was greatly augmented in size and gorged with blood. The right lung was congested and adherent to wall of thorax; considerable effusion in this pleural sac. The cæcum and upper portion of colon covered with deep blackish ulcers. A few small ulcers in ileum.

No.	First inoculation May 27.	Second inoculation June 10.	Exposure in infected pen.	Time of death.	Days after first exposure.
	c. c.	c. c.			
239	½	....	....	June 2†	....
242	½	2	June 24	July 17	23
244	½	2	...do ...	July 9	15
245	½	2	...do ...	....	....
*243	....	....	...do ...	July 13	19
240	½	3	...do ...	July 10	16
254	½	3	...do ...	...do ...	16
255	½	3	...do ...	July 20	26
256	½	3	...do ...	....	....
*253	....	....	....	....	....

* Check.   † From inoculation.

The foregoing experiments demonstrate the important fact that pigs can not be made insusceptible to hog cholera by subcutaneous injections of pure cultures of hog cholera bacteria, according to the method used by Pateur in anthrax vaccination. They show how injudicious it is to draw conclusions extending to infectious diseases in general from experiments made upon one disease only.

THE EFFECT OF FEEDING SMALL QUANTITIES OF CULTURES.

It has been already demonstrated (p. 107) that hog cholera can be produced in pigs by feeding such as had been starved for a day with considerable quantities of cultures of the specific bacteria. From 200 to 400 cubic centimeters of a beef-infusion culture were necessary to destroy the animal.

Assuming that small doses may produce a mild form of the disease,
it seemed worth while to determine whether animals which had been
fed with small quantities of culture material would resist infection when
placed with sick animals in thoroughly infected pens. The following
experiment shows that even after this treatment pigs are still susceptible to the disease:

Nos. 379, 380, 381, 382, and 370 were starved for thirty-six hours and
then fed pure cultures of hog cholera bacteria in simple beef infusion.
The quantities given are tabulated below. They were diluted with
ʾbeef broth to facilitate feeding.

No.	Age, January 21.	Quantity of culture liquid fed.	Placed in infected pen.	Died in infected pen.	No. of days after exposure.
		c. c.			
380	5 months	10	Apr. 15	June 20	66
379	....do	20	...do	May 24	39
382	....do	40	...do	May 10	25
381	...do	60	...do	May 31	46
370	....do	100	...do	Apr. 20*	5
†383	4 months	....	Apr. 19	May 11	22
†387	....do	....	...do	Apr. 30	11
†394	....do	....	Apr. 30	May 17	17
†399	....do	....	.. do	....	....

* From injury.   † Checks.

All appeared to be affected by the feeding on the following day by
refusing food more or less. Slight improvement on the second day, with
the exception of 381. On the third day 381 still very ill. On the fourth
day 382 seems more affected than 381. Within the next few days 382
refuses food, the rest eat. In the second week they had recovered.
The growth of 382 was materially checked by the feeding, so that it
had not gained any weight even after several months.

April 15 they were transferred to a pen infected with hog cholera, in
which acute cases had lately appeared. A number of check animals
were transferred to the same pen soon after. The result may be briefly
summed up as follows:

No. 379 began to fail from May 10 and died May 24. Lymphatic
glands generally enlarged and congested; those of meso-colon very
much so. Spleen very much enlarged; quite firm. Large number of
yellowish coagula in abdominal cavity. Adhesions between cæcum
and adjacent organs. Liver very much cirrhosed; gives a gritty sensation when cut. Several deep, extensive ulcers from 1 to 2 inches
across in cæcum, involving the serous membrane. In lower colon and
rectum considerable congestion and hemorrhage. Lungs normal, with
exception of a few subpleural ecchymoses. Both sides of heart filled
with white thrombi. This animal, therefore, had chronic hog cholera,
which, no doubt, caused the fatal peritonitis.

No. 382, the stunted pig, showed signs of the disease May 1 and died
May 10. Very emaciated; connective tissue of body stained deep yellow. Spleen very large and full of blood. Lymphatics but slightly
enlarged and reddened. Considerable yellow serum in abdominal

cavity. Right heart filled with dark and washed clot. Lungs normal, excepting a few subpleural ecchymoses. Extreme cirrhosis of liver. Mucosa of great curvature of stomach deeply congested, mucosa of large large intestine in patches.

No. 381 took sick soon after its removal to the infected pen, but the disease seemed to be mild until it was found dead May 31. Animal with a fair quantity of fat. Skin of ears, pubis, and throat deeply reddened. Lymphatic glands generally enlarged and infiltrated with blood. Spleen very much enlarged, but tissue quite firm. Lungs hypostatic; no hepatization. Epicardium covered with pale spots of extravasation. Liver partially sclerosed; hepatic ducts filled with viscid yellow bile. Medulla of kidneys dotted with petecchiæ. Mucosa of large intestine apparently normal. In the lower colon the feces are blood-stained, and occasional small clots are loosely attached to mucosa.

No. 370 was injured by fighting after being transferred to the infected pen so as to be scarcely able to stand, and died April 20. On both hind limbs there was extensive extravasation in the subcutis. The superficial inguinal glands were deeply congested. Spleen enlarged and congested. A moderate number of bacteria present, which are not hog cholera bacteria as shown in cultures. Left lung collapsed, reddened in part and adherent to chest wall by a plastic exudate. Centers of acini of liver deep red, tissue soft. Kidneys congested. Stomach congested along fundus; digestive tract otherwise normal. This animal died from injury to limbs and lung, and there may have been septic infection as indicated by bacteria in the spleen.

No. 380 became ill May 6, died June 20. Decomposition advanced. Lungs and glands normal. Liver small, sclerosed. Superficial necrosis in cæcum and upper colon with cicatrices of healing ulcers.

Gelatine tube cultures from the spleens of 382, 383, and 397 showed innumerable colonies of hog cholera bacteria. Cultures from the spleen of 381 remained sterile. A larger number of cultures would probably have shown their presence.

As to the fate of the check pigs exposed with the preceding lot very little need be said.

No. 383 had been fed with 250 cubic centimeters of a beef-infusion culture of the hog cholera bacteria from Nebraska as far back as December 19. It recovered, however, after an illness of three or four days. March 16 it received a tracheal injection of swine plague bacteria, without any untoward result. It was exposed to hog cholera as a check April 19, and died very suddenly May 11. It had hemorrhage in subcutis, lungs, lymphatic glands, in stomach, small and large intestine. In the cæcum there was an ulcerated patch about 2 inches across. Its base was very firm and nearly one-third inch thick. Was the ulcer the result of the feeding five months ago?

No. 387 became very dull and off feed one week after its removal to the infected pen, and died April 30. Skin of abdomen reddened. Lymphatics generally hemorrhagic. Ecchymoses on lungs. Extravasations in medulla of kidneys. Beginning ulceration in large intestine. Congestion and slight hemorrhage in fundus of stomach.

No. 394 died May 17 after a few days of dullness. Hemorrhagic condition of spleen and lymphatic glands. Inflammation and extravasation in stomach. Ulceration and intense congestion of large intestine. Ecchymoses beneath pleura of lungs.

No. 399 was alive and apparently well June 15. Hog cholera bacteria were found in the spleen of 387 and 394, especially numerous in the latter. They were not looked for in 383.

There was no protection from the feeding of cultures even in those animals which were most severely affected thereby. There was, however, a marked difference between this lot and the check animals, in that the latter died of the most severe septicæmic or hemorrhagic type of the disease, while the former succumbed to a chronic infection. If we take the average duration of the disease in the pigs fed with cultures and those not fed, the ratio will be approximately as 19 to 8 or as $2\frac{1}{3}$ to 1. These figures indicate a considerable retardation of the fatal issue, which is equivalent to a partial immunity. Another fact of importance needs to be considered in connection with these figures. Three of the pigs fed with cultures were affected with a more or less advanced cirrhosis of the liver and the *post-mortem* notes will show that this liver disease largely contributed to the death of the animals. An important question arises in this connection as to the origin of the cirrhosis. It may have been due to the feeding of the cultures. If so, the remedy would be as bad as the disease. But even setting aside this possibility, the method of feeding living bacteria can only have a theoretical importance in leading to other methods. It not only scatters the living virus, but it may induce ulceration in the large intestine which, while not recognizable, may be slowly destroying the animal.

INJECTIONS OF STERILIZED CULTURE LIQUID TO PRODUCE IMMUNITY.

Bacteriological investigations of the past few years have shown pretty conclusively that during the multiplication of pathogenic bacteria there are formed chemical substances or ptomaines which are poisonous to the animal economy. The researches of Brieger have done more than any others to confirm this theory. It is now generally believed that it is these poisons which produce such grave symptoms in infectious diseases rather than any other vital manifestation of bacterial growth. In the spring and summer of 1886 Sirotinin made some experiments (*Zeitschrift für Hygiene*, I, 463) with the bacillus of typhoid fever, in which he endeavors to show that the smaller experimental animals can not be infected by this bacillus, but that death caused by injection, subcutaneously or otherwise, must be attributed to an intoxication caused by the presence of a ptomaine in the cultures. The results actually proved that the injection of sterilized cultures may produce death accompanied by lesions resembling those produced by living bacilli. Beumer and Peiper (*Zeitschrift für Hygiene*, II, p. 110), after a long series of experiments, are brought to the same conclusion, that the typhoid-fever organism does not multiply in the body of smaller experimental animals, that there is, indeed, no true infection, and that the severity of the symptoms depends entirely upon the quantity of culture material injected; in other words, upon the quantity of the poison or ptomaine therein contained. They also point out the important fact to which we wish to call attention, that death does not follow the injection of large doses if small non-lethal doses have been given previously, and from this fact they argue that perhaps immunity may finally be brought about by the injection of sterile cultures in successively larger doses. That the chemical products of bacterial growth may produce immunity is no new theory, but it seems to have gained ground but recently among investigators.

TESTS WITH STERILIZED CULTURES ON PIGEONS.

In our experiments pigeons were selected at first, because more easily obtainable and manageable. They proved, however, a very good choice, since they do not take the disease readily and thus are made insusceptible by small doses of culture liquid. A culture in beef infusion, containing 1 per cent. peptone, in which active multiplication of bacteria ceases in a few days, is fatal to adult pigeons, as a rule, when three-fourths of a cubic centimeter is injected subcutaneously over the pectoral muscle or superficially into the muscle itself. The intra-muscular injection is more rapidly fatal. The effect of these inoculations has already been dwelt upon, but it is restated here on account of the important principle involved.

The pigeon, after such an injection, may be dead within twenty-four hours. The inoculated pectoral muscle is more or less discolored throughout its depth. There may be a regurgitation of food from the crop, as grains are found mixed with mucus in the mouth and œsophagus. The injected bacteria are present in small number in heart's blood, liver, and spleen.

About one-half of the pigeons do not die so soon. The bird stands quietly in a corner of the coop, with feathers ruffled, wings slightly separated, and tail feathers drooping. The discharges are usually abnormally liquid, at times mixed with considerable mucus. The bird usually dies within a week. The pectoral muscle will then be found extensively necrosed, the surrounding tissue very hyperæmic. The injected bacteria are found, sometimes in considerable number, in the liver and heart's blood.

The rapidly fatal cases might be regarded as the result of simple intoxication or ptomaine poisoning. There is, however, some bacterial multiplication. In the more chronic cases there is an undoubted infection, characterized by multiplication of bacteria in the internal organs. Pigeons are far more susceptible in winter than in summer, consequently in the heat of midsummer the control animals occasionally resist and thus impair the value of the experiments.

The experiments were carried out as follows:

Culture tubes containing about 10 cubic centimeters of beef infusion with 1 per cent. peptone were inoculated with hog cholera bacteria and placed in the incubator at $34°-36°$ C. After a certain number of days, varying from three to ten, the tubes were exposed to a temperature of $58°-60°$ C. for about one hour. Inoculation of fresh tubes showed that bacteria had been destroyed. This test was always resorted to to make sure that no living bacteria were injected.

From 1 to 1.5 cubic centimeters of this culture liquid were injected with a hypodermic syringe beneath the skin of one pectoral muscle. This injection was repeated once or twice. Some days after the last injection the bird was inoculated with living bacteria. About ¾ cubic centimeter of a beef infusion peptone culture was injected beneath the skin of the other pectoral or into the superficial layer of muscular fibers. The vaccinated pigeons remained alive and well; the control pigeons nearly all died. These statements are best illustrated by the tabulated results of a few experiments.

The first one, made in January, 1886, is given below. The control bird and the one which had received a very small quantity of sterile culture liquid died within two days after the test inoculation; the rest were well more than a week later.

I.—1885–'86.

No.	Injection of—					Culture liquid containing living bacteria.	Remarks.
	Sterilized culture liquid.						
	Dec. 24.	Jan. 21.	Jan. 29.	Feb. 6.	Total.	Feb. 13.	
	c. c.	c. c.	c. c.	c. c.	c. c.	c. c.	
10	.4	1.5	1.5	1.5	4.9	¾	Well February 20.
11	....	1.5	1.5	1.5	4.5	¾	Do.
12	....	1.5	1.5	.1.5	4.5	¾	Do.
13	....	1.5	....	1.5	3.	¾	Do.
8	.8	....	....	....	.8	¾	Died in forty-eight hours.
14	....	....	....	....	....	¾	Died in twenty-four hours.

A second series of injections, made to confirm these rather remarkable results, was equally unequivocal in its answer:

II.—1886.

No.	Subcutaneous injection of—				Fresh culture liquid.	Remarks.
	Sterilized culture liquid.					
	Feb. 19.	Feb. 24.	Mar. 2.	Total.	Mar. 8.	
	c. c.	c. c.	c. c.	c. c.	c. c.	
16	1	1	¾	2¾	¾	Well after several weeks.
17	1	1	1	3	¾	Do.
18	1	1	1	3	¾	Do.
19	1	....	1	2	¾	Do.
20	1	....	1	2	¾	Do.
21	1	....	1	2	¾	Do.
22	....	....	....	....	.....	Dead March 9.
23	....	....	....	....	¾	Do.
24	....	....	....	....	¾	Remained well.

The third control bird (No. 24) was of a different race of pigeons. A good authority consulted at the time regarded it as having some of the characters of the carrier pigeon. Leaving this aside, the result is sufficiently convincing.

In a series of experiments made more recently, cultures were used which had been concentrated by evaporation *in vacuo*. About 100 cubic centimeters of beef infusion, containing 2 per cent. peptone and one-half per cent. sodium chloride, was inoculated and kept in the incubator for five days. It was then reduced to 20 cubic centimeters by evaporation at 40° C., and sterilized at 60° C., for three-quarters of an hour. Subsequent inoculation of fresh tubes showed that the liquid was free from living bacteria. The injections were made as usual, the needle entering the pectoral muscle very superficially.

III.—1887.

No.	Sterilized concentrated culture liquid.			Fresh culture liquid.	Remarks.
	Apr. 19.	Apr. 22.	Equivalent of ordinary culture fluid.	Apr. 25.	
	c. c.	*c. c.*	*c. c.*	*c. c.*	
1	$\frac{1}{4}$	$\frac{1}{4}$	$2\frac{1}{2}$	$\frac{3}{4}$	Well May 31.
2	$\frac{1}{4}$	$\frac{1}{4}$	$2\frac{1}{2}$	$\frac{3}{4}$	Do.
3	$\frac{1}{4}$	$\frac{1}{4}$	$2\frac{1}{2}$	$\frac{3}{4}$	Slightly ill April 26; well May 31.
4	1	....	5	$\frac{3}{4}$	Well May 31.
5	1	....	5	$\frac{3}{4}$	Do.
6	1	....	5	$\frac{3}{4}$	Do
7	....	....	....	$\frac{3}{4}$	Died April 26, 9 a. m.
8	....	....	....	$\frac{3}{4}$	Died April 26, 1 p. m.
9	....	....	....	$\frac{3}{4}$	Died April 30.

In order to determine whether the introduction of sterilized culture liquid into the digestive tract of pigeons would be equally efficient in giving them immunity, three pigeons were fed on three days (September 8, 10, 13), each receiving about 30 cubic centimeters in all. The culture liquid consisted of beef infusion, to which 1 per cent. peptone had been added. The feeding was carried out by introducing a catheter into the pigeon's crop and injecting through this by means of a syringe. On September 16, three days after the last feeding, these three pigeons and two checks were inoculated with $\frac{3}{4}$ cubic centimeter of unattenuated culture liquid. The three fed pigeons and one check died from the effects of the inoculation, the second check was well September 29. The blood and liver of No. 1 were examined by adding a few loops of heart's blood and a minute piece of liver tissue to two roll-cultures. In each about 50 colonies developed. In the liver of No. 2 bacteria were very numerous. The feeding therefore had no effect whatever in conferring immunity.

Fed with sterilized cultures five days and nine days old.			Inoculation of unheated culture liquid Sept. 16.	Remarks.
No.	Sept. 8	Sept. 10	Sept. 13.	
	c. c.	c. c.	c. c.	c. c.	
1	10	10	10	?	Dead September 17.
2	10	10	10	?	Dead September 23.
3	10	10	10	?	Dead September 24.
4	....			?	Died September 19.
5	....			?	Well September 29.

The protected pigeons very rarely show any signs of illness. They are as active and as eager for food as before the final inoculation. In all a small sequestrum is formed in the pectoral muscle, where the injection has been made, and this, at the end of a few weeks, is surrounded by a dense membrane which seems to act as an absorbing surface for the sequestrum. No indication of any reaction is found on the side into which the sterile cultures are injected. The three tables may be summarized as follows: Of twenty-four pigeons sixteen received sterilized cultures, eight being reserved as checks. Of the former none succumbed to the final inoculation; of the latter seven, or 87.5 per cent.

The conclusion to be drawn from these experiments is obvious. The birds are protected by the injection of sterilized cultures so as to resist a fatal dose of living bacteria. The sterilized cultures contain only the products of bacterial growth. Among these the ptomaine-like bodies—some of which we now know, owing to the researches of Brieger—are very likely the agents that produce immunity.

In the pigeon the mode of infection before and after vaccination is probably as follows: The injected bacteria multiply very actively in the muscular tissue; the ptomaine there produced may enter the circulation in quantities large enough to produce speedy death. If the animal resists for a time, the absorbed ptomaine reduces the vitality of the tissues to such a degree that bacteria, entering the circulation, begin to multiply in the internal organs. The additional quantity of ptomaine thus produced finally kills the bird.

When ptomaines in culture liquids have been previously introduced the first shock caused by the local production of ptomaines in the muscular tissue is overcome. The bird resists successfully general infection until the bacteria have been destroyed locally. The process is then checked, and the sequestrum in the muscular tissue becomes encysted.

It has already been stated (p. 69) that rabbits, mice, and guinea-pigs are very susceptible to the inoculation disease, a millionth of a cubic centimeter of culture liquid being sufficient to cause a fatal issue. Experiments like those on pigeons were made upon rabbits with slightly larger doses, but no immunity was gained thereby. The rabbits succumbed invariably after inoculation. (See report of Department Agriculture for 1886, p.

637.) In the light of later experiments* made in Pasteur's laboratory, the failure of the method in case of rabbits must simply be ascribed to the small quantity of culture liquid used. In the experiments there recorded from 80 to 120 cubic centimeters of culture liquid were required to protect guinea-pigs from malignant œdéma, while in our rabbit experiments only 5 to 6 cubic centimeters were used in all. We did not delay, however, with the further demonstration of this principle upon small animals, and applied it experimentally to pigs. The quantities injected were not sufficient, however, and the animals took the disease, as the following experiments show:

TESTS WITH STERILIZED CULTURES ON PIGS.

Experiment 9.—In order to test the effect of heated cultures upon pigs, the following experiments were made March 1: Two animals (Nos. 162 and 173) received hypodermically each 9 cubic centimeters of a second and third culture, twelve and thirteen days old, respectively, which had been devitalized by heat. March 9 a second dose of 9 cubic centimeters was given in the same way, using a fifth and eighth culture eighteen and fourteen days old, respectively. These cultures were made in beef infusion containing 1 per cent. peptone, excepting one, which contained about 2 per cent. of blood serum in place of the peptone. After the second inoculation of No. 162 a swelling appeared on one side. Both were fed with viscera infected with hog cholera, and placed with sick and dying pigs in a large infected pen. No. 162 was found dead March 29, and No. 173 April 5. The appended table and notes give a summary of the experiment:

No.	Mar. 1.	Mar. 9.	Total.	Fed.	Died.	No. of days after feeding.
	c. c.	c. c.	c. c.			
162	9	9	18	Mar. 19	Mar. 29	10
173	9	9	18	Mar. 19	Apr. 5	17

No. 162. Subcutaneous fatty tissue much reddened. Mucous membrane of stomach considerably ulcerated; mucous membrane of small intestine deeply congested. For 8 or 10 feet above the ileo-cæcal valve the mucous membrane of ileum is completely necrosed. Large ulcers in cæcum and upper portion of colon.

No. 173. Subcutaneous fatty tissue slightly reddened. Petecchiæ under pulmonary pleura. Extravasations under serosa of cæcum and colon. Inflammatory adhesions of large intestine with walls of abdomen. A patch of extravasation beneath peritoneal layer of dorsal abdominal wall nearly 2 inches across. Spleen very much enlarged and softened. The mucous membrane of large intestine and several feet of ileum necrosed and breaking down. Fundus of stomach deeply congested.

This experiment clearly showed that this method was no protection to the animal when the latter was infected by feeding.

---

*Annales de l'Institut Pasteur, December, 1887, January, 1888.

It now became necessary to determine whether this method would confer immunity upon animals simply exposed to the disease by cohabiting with diseased animals in infected pens. Observations made upon other diseases by investigators, and by us upon this disease, seem to lead to the inference that it frequently depends on the quantity of virus introduced into the system whether the disease will make its appearance or not. In feeding, this quantity is considerable; in simple exposure in infected pens to diseased pigs, the amount of virus taken into the body with the food and drink is necessarily in small and repeated doses. The following was therefore planned:

Experiment 10.—Four pigs (Nos. 163, 164, 177, and 196) were inocu. lated March 13 with heated virus, each receiving 4½ cubic centimeters beneath the skin of each thigh. The cultures in beef infusion with 1 per cent. peptone were about fifteen days old when heated. The second inoculation was made March 16 from a culture in an Erlenmeyer flask about eleven days old, and containing about 50 cubic centimeters of culture liquid. Each animal received 10 cubic centimeters as before.

March 31 these animals, together with two check pigs (Nos. 195 and 201), were placed in a large infected pen. Within a period of three weeks from this date at least fifteen pigs died of hog cholera in this pen. The two check animals died on the 14th and 19th of April, respectively. Of four vaccinated animals only No. 163 showed signs of the disease and gradually developed into a chronic case, dying of general debility on May 1. The three other vaccinated animals remained apparently well for months after, although constantly exposed to the disease in the infected pen.

No.	Vaccination. Mar. 13	Vaccination. Mar. 16	Exposure.	Died.	Number days after exposure.
	c. c.	c. c.			
163	9	10	Mar. 31	May 1	Thirty-one days.
164	9	10	....	....	
177	9	10	Mar. 31	July 23	Three months and twenty-three days.
196	9	10	...do	July 7	Three months and seven days.
195	....	....	...do	Apr. 19	Nineteen days.
201	....	....	...do	Apr. 14	Fourteen days.

Autopsy notes.—No. 163. Spleen not much enlarged; texture firm; effusion into pericardial and thoracic cavity; lymphatic glands enlarged but pale; two ulcers in stomach; small intestine normal; mucosa of cæcum and colon studded with many extensive and deep, yellowish ulcerations. On cover-glass preparations of the spleen only a few bacteria could be seen. Two liquid cultures inoculated from the same organ remained sterile. No colonies appeared in the gelatine tube inoculated with blood from the heart. A few developed in the gelatine tube inoculated from the spleen.

No. 195. Spleen greatly enlarged; gorged with blood; very friable; shreds of a fibrinous exudate on serosa of intestines; much serum in abdominal cavity; petecchiæ on epicardium of auricles; small ventral lobes of lungs hepatized; mucous membrane of gall bladder ulcerated;

extensive ulceration and inflammation of mucous membrane of cæcum and colon. Hemorrhagic inflammation of kidneys.

No. 201. Spleen but slightly enlarged; lungs extensively hepatized; intense congestion and commencing ulceration of the mucosa of large intestine; stomach and portion of ileum similarly congested. Though no bacteria were found on a cover-glass preparation, a pure culture was obtained by carefully dropping a piece of spleen tissue into a culture tube. This was tested on gelatine.

After apparently resisting the infection for several months, the remaining pigs (Nos. 164, 177, and 196) were transferred to a clean pen. No. 177, not very thrifty, began to decline, and finally died July 23. Among the most prominent lesions were a plastic exudate on the epicardium and numerous large old ulcers in the large intestine. The mucosa itself was extensively pigmented. No. 196, on removal from the infected pen, seemed in good condition, but it died July 7, after some days of unthrifty condition. In this case, the mucous membrane of the large intestine was pigmented and there were what appeared to be cicatrices of old ulcerations. In all of the large serous cavities there was considerable effusion. In cover-glass preparations of the spleen there were no hog cholera bacteria to be seen, but numerous bacilli resembling those of malignant œdema.

Experiment 11 was made in the same way upon Nos. 197 to 200, inclusive, and No. 157. March 24 each animal received in the thigh about 10 cubic centimeters of a mixture of heated cultures in beef infusion with 1 per cent. peptone about fourteen days old. March 29 an equal amount was injected, one-half into each axilla, these cultures being about fifteen days old. These animals were kept until April 20, when all but No. 157 were placed in the large infected pen. From that date on pigs died almost every day of the disease, so that the infection must have been quite thorough. Unfortunately no check animal was exposed at the same time. In these animals the slight swelling at the seat of inoculation disappeared in a few weeks.

They remained well, with the exception of No. 199, which became emaciated and was found dead May 19, about one month after exposure. The three remaining animals were apparently unaffected nearly two months after exposure. At this time No. 197, which appeared rather thin, was killed, to determine if any ulcerations were present. But the mucous membrane of the intestine was entirely normal, with no indications of former ulcerations.

No.	Injection of heated virus.		Time of exposure	Remarks.
	Mar. 24	Mar. 29		
	c. c.	c. c.		
197	10	10	Apr. 20	Killed June 10; healthy.
198	10	10	...do ...	Well June 10.
199	10	10	...do	Died May 19.
200	10	10	...do ...	Died July 12.
157	10	10	May 25	Died June 28.

Autopsy.—No. 199. Slight extravasation in subcutaneous connective tissue. Spleen somewhat enlarged, filled with blood, friable; considerable effusion in peritoneal cavity. Right lung in part hepatized; pleuritic adhesions to chest-wall; hemorrhage in and about pelvis of kidneys; lymphatic glands purplish; extensive and deep ulceration of the mucosa of large intestine.

Pig 197, killed for examination, was very anæmic. There was some pale serum in abdominal cavity. The kidneys and lymphatic glands showed evidence of chronic inflammation. The lungs were exceedingly pale. No evidence of inflammation or ulceration in any portion of the intestinal tract.

It must be borne in mind that these animals were constantly exposed for a period of several months to the virus of the disease, and that a continual struggle between the organism and the invading parasites must have been going on, which naturally would tend to lower the vitality. Such severe conditions as these are probably never realized among herds.

The later history of No. 200 does not, however, bear out the first supposition that complete immunity was attained. After being continually exposed in the infected pen from April 20 to June 21 it was removed to a clean pen, where it continued to grow very weak. It died July 12. The autopsy revealed a plastic pleurisy over the right lung and a fibrinous exudate upon the epicardium. The mucosa of the cæcum was extensively necrosed; in the colon the ulcers were isolated; the solitary follicles were very prominent. A small bit from the epicardial exudate was placed beneath the skin of two mice. One of them died on the eighteenth day. The spleen was greatly enlarged. Numerous hog cholera bacteria were present in this organ and the liver. The epicardial exudate of the pig must have contained but very few, for they could not be demonstrated in cover-glass preparations. The long period of time from the inoculation of the mouse to its death is also evidence of a very small quantity of virus.

No. 157, inoculated with the rest, became quite lame in the hind limbs, so that it was thought best not to expose it to the disease in the infected pen for the time being. It soon recovered its power of locomotion, and was transferred to the infected pen May 25 and removed therefrom June 28. In the new pen it grew rapidly weaker, and died June 28. On *post-mortem* examination the right lung was found entirely hepatized and adherent to the chest-wall. The mucosa of cæcum and colon was studded with large and deep ulcers; that of the fundus of stomach was deeply congested.

Experiment 12.—It became desirable to determine whether repeated subcutaneous injections of heated cultures, until a large amount had been introduced into the system, would be more efficacious in producing immunity. For this purpose the culture liquids were concentrated, by using a 2 per cent. solution of meat extract with 2 per cent. peptone for some of the injections; for the remainder a 2 per cent. solution of peptone in beef infusion. The cultures were made in Erlenmeyer flasks, plugged with cotton wool.

The table given below gives the date of the injection and the quantity used each time. It will be noted that Nos. 191 and 194 received two, Nos. 216 and 218 three, and Nos. 217 and 221 four doses each of the heated culture liquid. The injections were made two days apart, the exposure in the infected pen and among diseased animals about one week after the last inoculation. Nos. 220, 232, and 235 were placed in the infected-pen at the same time, to determine the virulence of the infection upon pigs which had not received any injection.

No.	Heated virus.				Total.	Exposure in infected pens.	Remarks.
	Apr. 20.	Apr. 22.	Apr. 24.	Apr. 26.			
	c. c.	c. c.	c. c.	c. c.	c. c.		
216	0	8	7½	....	24½	May 4...	Died May 17.
217	9	8	7½	8	32½	...do	Died May 19.
218	9	8	7½	....	24½	...do	Do.
221	9	8	7½	8	32½	...do	Died May 23.
191	....	10	8	18	...do	Do.	
194	....	10	8	18	...do	Died May 19.	
*220	....	....	....	....	....	...do	Died May 17.
*232	....	....	....	....	....	...do	Died May 23,
*235	....	....	....	....	....	...do ...	Died June 12.

* Checks.

All of the inoculated and control animals died with periods ranging from thirteen to nineteen days, only one living thirty-nine days, and this one a control animal. Of those that had received two doses, No. 191 died May 23 (nineteen days after exposure) with considerable ulceration in cæcum and colon. No. 194 died May 19, with extensive and deep congestion of the lymphatic glands in general, of the kidneys, stomach, and large intestine. In the latter, ulceration was not yet begun. No. 216, which had received three doses, died very unexpectedly thirteen days after exposure. The lesions were of the hemorrhagic type, involving extravasations and ecchymoses of the intestinal tract, more especially of the large intestine, heart, lungs, lymphatic and subcutaneous fatty tissue. Ulceration in large intestine very slight, the congestion being intense. No. 218, treated like the former, died fifteen days after exposure. The lesions were like those of No. 216, but not so severe. Ulceration as yet very slight.

Nos. 217 and 221, which had received four injections, died fifteen and nineteen days after exposure, respectively. The lesions in No. 217, which died very suddenly, were hemorrhagic in character, the ulceration in the cæcum and colon being quite superficial. In No. 221 the ulceration was more pronounced, the general congestion and extravasation much less so.

Of the control animals the lesions of No. 220 were of the hemorrhagic type, resembling those of No. 194 very closely. In No. 232 there was extensive ulceration of the mucous membrane of the large intestine. In No. 235, which lived for thirty-nine days after exposure, the mucosa of the cæcum and upper portion of the colon was involved in complete necrosis nearly 5 millimeters thick. Beyond this the necrosis took the form of isolated ulcers. Owing to the depth of the ulceration inflammatory adhesions had formed between the cæcum and adjacent organs. There was no reactionary swelling of the inoculated animals at the point of injection.

Those animals in which the disease took the hemorrhagic type succumbed very suddenly, as if the invasion had taken place in a single day. In those animals in which symptoms of weakness and loss of appetite appeared some days before death the well-defined lesions were as a rule limited to the large intestine in the form of ulcerations. The former cases represent a class in which the bacilli invade the entire vascular system; in the latter the absence of a general congestion and extravasation seems to indicate a more local multiplication of the specific disease germs in the intestinal tract.

This mode of vaccination, as shown by the results recorded, did not prove to be any protection to the animals, as they died, most of them, within a brief period after exposure from a very acute attack of the disease.

The spleen examined in about one-half of these cases contained the bacillus of hog cholera, usually in large numbers. From a few cultures were made which were found pure.

Experiment 13.—A final experiment was tried in which each animal received hypodermically 40 cubic centimeters of heated culture liquid in two doses. The cultures were made in beef infusion with 1 per cent. peptone, the growth being killed by a temperature of 58° C. the third day after inoculation. The flasks used were shaped like Erlenmeyer flasks, a glass cap being fitted over the flask by means of a ground-glass joint, which contracted into a straight narrow tube plugged with glass wool. The removal of a cotton-wool plug was thus avoided, the cap being removed for inoculation. This culture flask affords better ventilation and a more rapid evaporation of the culture liquid than does the culture tube with the bent ventilating tube.

The following table gives all the facts necessary for an understanding of the experiment and its results:

No.	Heated virus.		Total.	Exposure in infected pens.	Remarks.	Days after first exposure.
	June 14.	June 17.				
	c. c.	c. c.	c. c.			
231	20	20	40	June 21	Died July 7 .	16
233	13	20	33	do	Died July 9..	18
266	20	20	40	do ...	.. do	18
*230	....		......	. do	Died July 8 ..	17
267	20	20	40	.. do	Died July 6 .	15
268	20	20	40	. do ...	Died July 10	19
269	20	20	40	... do ...	.. do	19
*270	....	....	....	...do	Died July 9...	18

* Checks.

It will be seen that all the experimental animals died, inoculated as well as check animals, within a few days of one another, death taking place about sixteen to eighteen days after the first day of exposure. A brief synopsis of the *post mortem* appearances will not be amiss in this connection.

In No. 231 the spleen was very much enlarged and gorged with blood. The intensely congested mucous membrane of the cæcum and colon was dotted with small superficial ulcerations. In No. 233 the congestion of spleen, and ulceration with congestion of the large intestine, were also marked. No 266 presented the same lesions. The ulcers in the cæcum were from one-quarter to one-half inch across No. 230 (check) differed from the preceding cases in presenting severer lesions, greatly enlarged and congested spleen and lymphatic glands, superficial necrosis of the entire colon and great thickening of the walls, extensive extravasation of blood beneath the mucosa of duodenum.

Of the second lot of four, treated in the same way, No. 267 presented very severe lesions, consisting of intense congestion and extravasation, involving the spleen, lymphatic glands, lungs, and kidney. The left lung was adherent to costal pleura. There was considerable hemorrhage in the pelvis of both kidneys. The large intestine was least changed, the mucosa being slightly ulcerated and containing some hemorrhagic spots and points. This animal was first to die out of this lot of eight. In No. 268 the congestion involved the lymphatic glands generally and the mucosa of the large intestine, which was extensively necrosed in its upper portion. No. 269 resembled No. 267 in the severity of the lesions. The lungs were not affected, however, while the ulceration of cæcum and upper colon was very extensive and deep. No. 270 (check) presented extensive ulceration of the large intestine and a greatly enlarged spleen. In five cases the spleen contained the bacteria of hog cholera more or less abundantly. In two none could be seen on one or two cover-glasses. No local swelling had developed at the points of injection in any of the inoculated animals.

The quantity of culture liquid in these cases was likewise too small to have any beneficial effect. Experiments are now being prosecuted in the same direction by injecting very large quantities. These are, how-ever, not completed.

The experiments on the use of sterilized cultures have thus not be-come practically applicable as yet. The great importance of this prin-ciple, however, has been duly appreciated within the past year in foreign countries, notably in France, where experimentation of this kind has been carried on by pupils and assistants of Pasteur. The principle is destined to be serviceable in other infectious diseases, and may eventu-ally take the place of the older methods of Pasteur which utilize attenu-ated cultures of bacteria. These are often sufficiently virulent to destroy sheep (anthrax), and in this way may themselves become the agents that distribute the disease. For this reason anthrax vaccination is only practiced in those regions where the disease is more or less endemic, and where the losses exceed a certain percentage each year.

Chantemesse and Widal* have recently endeavored to set aside the claim of the Bureau of having first demonstrated this principle, in favor of the more recent researches made in Pasteur's laboratory. Hüppe† has, however, shown the absurdity of this claim in pointing to the ex-periments of the Bureau as having been published nearly two years prior to those of Chamberland and Roux.

---

* *Annales de l'Institut Pasteur,* February, 1888.
† *Fortschritte der Medicin,* 1888, 289.

EXPERIMENTS ON THE ATTENUATION OF HOG CHOLERA BACILLI BY HEAT.

Heat has been used by Pasteur in the attenuation of anthrax virus by exposing cultures of anthrax bacilli to a temperature of 42°–43° C., continuously for a certain number of days. Cultures kept in a thermostat at this temperature for about thirty-one days were so attenuated that they were incapable of destroying animals larger than very young mice. Kept in the same conditions for only twelve days inoculation failed to destroy adult guinea-pigs.* The former culture was denominated the first, the latter the second vaccine. To make sheep immune they were inoculated subcutaneously with the first vaccine, twelve days later with the second vaccine. Subsequent inoculation with strong virus had no effect upon the vaccinated animals, although it was quite invariably fatal to those which had not been so treated.

Although Pasteur's discovery must be considered a scientific event of great importance, its practical application is by no means a perfect success. Experiments conducted by Koch, Gaffky and Löffler, in Berlin, have shown that the process of attenuation does not always go on uniformly, and that the strength of the vaccine can not always be relied upon. A few animals may die as a result of the first or second inoculation. This fact induced the last International Congress of Hygiene at Vienna to adopt the resolution that anthrax vaccination should not be practiced upon sheep in any locality unless the disease causes a loss of 2 to 3 per cent. annually. It was also shown in the experiments at Berlin that immunity after vaccination is not absolutely perfect when the virus is introduced with the food. This is perhaps the most common way in which infection takes place.

The results obtained by Pasteur are sufficiently valuable to make it at least desirable to try heat attenuation for other bacterial organisms, although it does not follow by any means that the same process will suffice for all or even a small number of disease germs, for these differ among one another very widely.

Kitt* has tried heat in the attenuation of the virus of Black Quarter in Germany by exposing the diseased muscular tissue, which had been thoroughly dried in the air and ground to a fine powder, to the steam of boiling water at 100° C. for four, five, and six hours continuously. The spores of the bacilli of this disease were sufficiently attenuated after a six-hours' exposure that sheep inoculated with the powder in certain doses remained well after inoculation with strong virus. The reaction

* Compt. rend. Acad. des Sciences March 21, 1881.
* Centralbl. f. Bakteriologie und Parasitenkunde, 1888, i. 571.

after the vaccinal inoculation was very slight. Hog cholera virus is destroyed by a fifteen to twenty minutes' exposure in a water-bath at 58° C.; a momentary contact with boiling water is sufficient to destroy it, so that Kitt's method is not applicable to it, but only to bacteria which form spores.

The following experiments were undertaken with a view to test the method of Pasteur on hog cholera bacilli, and to obtain, if possible, a vaccine similar to that employed in the prevention of anthrax. Although only a preliminary step has been taken in this matter, and the promise of favorable results is not flattering, we consider it best to publish the results thus far obtained.

The first step in the process was to obtain, if possible, a cultivation which should prove harmless to rabbits. This was to be accomplished by placing tubes in a thermostat, kept at a certain fixed temperature as nearly as possible, and inoculating rabbits from time to time to determine any diminution in the pathogenic activity.

April 9, 1888.—Four Salmon tubes of beef infusion peptone were inoculated from an *agar-agar* culture of hog cholera bacilli two weeks old and placed in a d'Arsonval thermostat, the internal temperature of which registered between 42° and 43° C. Two series of inoculations were made on rabbits, one from one of the original liquid cultures at different intervals, the other from a culture renewed at the end of every five days by inoculating a fresh tube. The result is most easily understood by examining the following table:

Agar culture.

		Apr. 9. b. i. p. *(a).	
		Apr. 14, b. i. p. (a₁).	
Apr. 19, rabbit inoculated		Apr. 19. b. i. p. (a₂).	Apr. 19, rabbit inoculated.
Apr. 25, died.			Apr. 24, died.
		Apr. 24, b. i. p. (a₃)	
Apr. 30, rabbit.		Apr. 30. b. i. p. (a₁).	Apr. 30. rabbit.
May 4, died.			May 5, died.
		May 4, b. i. p. (a₅).	
May 9, rabbit.		May 9. b. i. p. (a₆).	May 9, rabbit.
May 18, died.			May 16, died.

* Beef infusion plus 1 per cent. peptone.

The first inoculations were made April 19, with culture *a*, which had been in the thermostat ten days, and culture *a₁*, which had been freshly made on April 14. Both rabbits received about ½ cubic centimeter of the culture liquid hypodermically. Rabbit *a₁* died in five days with extensive coagulation necrosis in liver and enlarged spleen. Rabbit *a* died on the following day with the same lesions. In both hog cholera bacilli were very numerous on cover-glass preparations of spleen, and obtained pure in cultures. The same results were obtained in all subsequent cases, so that no further mention need be made of this.

Two rabbits were inoculated in the same way April 30, one from the original culture *a*, the other from *a₁*. Both died May 4 and 5, respectively. Lesions the same as with the first pair.

Two rabbits were inoculated May 9, one from the culture *a*, now thirty days old; the other from *a₅*, the fifth renewal of *a*. Both rabbits died on the 7th and the 9th day, respectively. The lesions were practically the same, with the addition of slight hemorrhagic lesions in the intestinal tract. The period of the disease was slightly prolonged.

The result was not very satisfactory. If any attenuation was going on, its final attainment would demand too long a period of time. The experiment was therefore stopped and another undertaken. The temperature of the thermostat was raised to 45° C., to hasten the process of attenuation.

Apr. 23, *agar* culture.

May 12, b. i. p. (b).

May 22, rabbit inoculated.   May 17, b. i. p. (b¹).

May 29, died.

May 22, b. i. p. (b²). .

It was found that cultures inoculated from *b* failed to develop at the assigned temperature, so that this experiment was not continued any farther. It deserves to be mentioned, however, that a rabbit inoculated from the original culture, which had been kept at 45° C. for ten days, died seven days after inoculation with enlarged spleen and coagulation necrosis in liver.

A temperature of $43\frac{1}{2}$–44° C. was next chosen and the experiment conducted in the same manner, as the appended table will show :

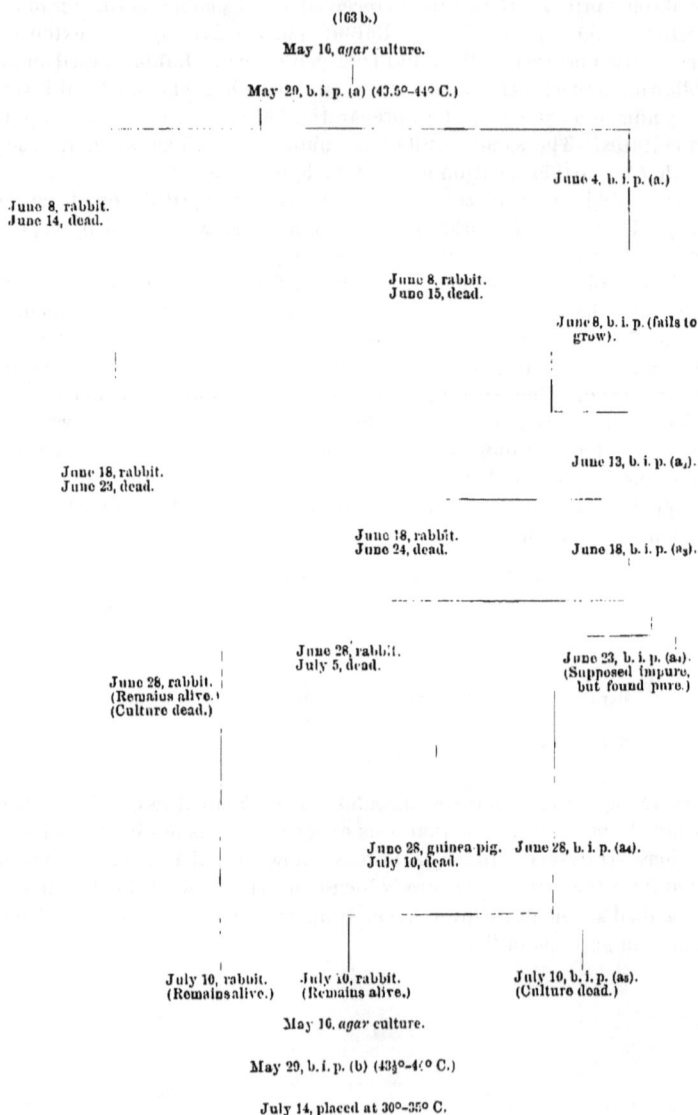

(163 b.)

May 16, *agar* culture.

May 29, b. i. p. (a) (43.5°–44° C.)

June 4, b. i. p. (a.)

June 8, rabbit.
June 14, dead.

June 8, rabbit.
June 15, dead.

June 8, b. i. p. (fails to grow).

June 18, rabbit.
June 23, dead.

June 13, b. i. p. (a₁).

June 18, rabbit.
June 24, dead.

June 18, b. i. p. (a₃).

June 28, rabbit.
July 5, dead.

June 28, rabbit.
(Remains alive.)
(Culture dead.)

June 23, b. i. p. (a₂).
(Supposed impure,
but found pure.)

June 28, guinea-pig.
July 10, dead.

June 28, b. i. p. (a₄).

July 10, rabbit.
(Remains alive.)

July 10, rabbit.
(Remains alive.)

July 10, b. i. p. (a₅).
(Culture dead.)

May 16, *agar* culture.

May 29, b. i. p. (b) (43½°–44° C.)

July 14, placed at 30°–35° C.

The rabbits inoculated from the original culture after remaining at the temperature of 43.5°–44° C. for ten and twenty days died from the inoculation disease, but those inoculated from the same tube. after thirty and forty-three days, remained permanently well. This was not due to an attenuation of the culture, but to its death. Turning to the series of rabbits inoculated from the cultures renewed every five or ten days, those receiving culture liquid ten, twenty, and thirty days old died from the inoculation disease, while one inoculated after forty-three days remained alive, because the culture was dead, *i. e.*, it failed to fertilize fresh tubes after repeated inoculation. An adult guinea pig, inoculated from the same culture material, thirty days old, died in twelve days as a result of the injection. In this case only a few drops had been injected. This experiment demonstrates that in general the pathogenic power of hog cholera bacilli is only destroyed by the death of the organisms themselves. This is a very important fact. It will be remembered that in the attenuation of anthrax bacilli their virulence was gradually diminished, and a time was reached when they failed to kill all but mice, while they still retained the power of multiplying in nutritive liquids. In the above experiments even guinea-pigs, which are less susceptible to this disease than rabbits, died twelve days before the culture was found dead. The latter may have been dead some days before this, for no tests were made meanwhile.

This fact has an important bearing upon the nature of the pathogenic activity of hog cholera bacilli. It shows that there are two elements involved, (1) the ptomaine action of the organisms; (2) their mechanical effect. That there is a ptomaine action of these bacilli has been conclusively proved in the experiments of the Bureau made upon pigeons several years ago. But this ptomaine action is evidently secondary to the mechanical effect of the bacilli in forming plugs or thrombi in the blood vessels and thus causing destruction of tissue by impeding the circulation. This tendency to act mechanically is not lost as long as the bacilli are alive, as shown by their fatal effect on rabbits and guinea-pigs shortly before they themselves are destroyed.

At the temperature employed (43.5°–44° C.), the original bouillon-peptone culture *a* died between the twentieth and the thirtieth day after the beginning of the exposure. The culture from this, renewed at the end of every fifth or tenth day, died between the thirtieth and forty-second day. Another culture *b* (see table), which had been removed from the thermostat after the forty-sixth day and kept at the temperature of the laboratory (about 30°–33° C., during July) was still fatal to a rabbit on the fifty-seventh day. Another rabbit, inoculated ten days later, remained well, and a fresh culture made at the same time remained sterile, showing that the apparent immunity of the rabbit was due to the death of the bacilli injected. This experiment also shows pretty conclusively that the pathogenic power of these specific organisms expires only with their life, and not long before.

It is evident from our own experiments, and more recent ones made in France and Germany, conducted on the same lines, that the amount of immunity which we may expect to gain from preventive inoculation will depend on the quantity of ptomaines produced by the specific microbes i. e., upon their poisonous nature. In other words, our success will depend upon the relation borne by the ptomaine to the disease process. If this factor is very great it is highly probable that preventive inoculation either with ptomaines or with attenuated cultures will be successful. But if there is in addition a mechanical element, which may or may not overshadow in importance the ptomaine element, the problem is not only complicated, but may fail.

There are two other points in connection with these experiments which demand attention. One is the variation in the length of life of the different cultures exposed to the same conditions. This would be a serious hindrance in obtaining vaccinal cultures of uniform strength, should this method ever prove successful.

There was a noticeable change in the appearance of the serial cultures after a sojourn in the thermostat. There was a tendency to multiply rather more abundantly, to grow in minute flakes and to rise to the surface to form a thin, unbroken membrane. The motility was somewhat impaired after a time. These changes gave rise to the suspicion of impurities, but tests on gelatine plates showed that the suspicion was unfounded. These experiments will be continued on pigs under similiar conditions to determine whether any immunity can be produced by this method.

HOG CHOLERA, OR DISEASES CLOSELY ALLIED TO IT, IN OTHER COUNTRIES.

GREAT BRITAIN.

This disease was first detected in 1862, where it has been known under various names, the most common of which is swine fever. Owing to the fact that there are recognized at the present time three distinct infections diseases of swine in different parts of Europe and America, it is impossible to state whether the disease known as swine fever in England represents two distinct maladies, hog cholera and swine plague, or hog cholera alone. This question can only be solved by a thorough bacteriological investigation, such as has not been made there in very recent years. There is enough evidence, however, to warrant the statement that hog cholera is prevalent in Great Britain. There is, on the other hand, no experimental basis for the statement that the other disease, swine plague, also exists there. These two diseases produce lesions of the large intestine, so easily confounded with each other, that great care must be taken in making a diagnosis. We must also consider that the three well-marked swine diseases now known as hog cholera, swine plague and *rouget*, were only a few years ago regarded as one disease designated by a great variety of names; that even to-day no clear idea exists among most pathologists as to what the precise differences between these diseases are.

In a brief report on swine fever in Great Britain, by Professor Brown, of the Agricultural Department, in 1886, we do not notice that any doubt exists in his mind as to the oneness of the disease in Great Britain and its identity with hog cholera. He even reproduces a plate from the report of the United States Department of Agriculture for 1885, in which the characteristic ulcers of hog cholera are shown as an illustration for his text.

That there may be several diseases included under the term swine fever is evident from the statements made in this report concerning the annual importations of diseased swine from other countries. Swine fever is said to have affected animals imported from Holland, Denmark, Sweden, Norway, and the United States. Now, hog cholera appeared in the North European countries for the first time in 1887. The swine marked diseased may have been affected either with swine plague, which exists on the Continent, or with lighter non-infectious forms of lung disease not uncommon in pigs at any time in the year.

173

We quote from Professor Brown's report those parts which may prove interesting to American readers, using the language of the report itself whenever possible:

Symptoms of swine fever.—It is most important that the farmer should be acquainted with the signs which indicate the existence of swine fever at the earliest period of its development; but unfortunately the disease is very difficult of detection in the early stage in the greater number of cases, and the symptoms which are generally believed to be specially indicative of the affection, viz., redness of the skin on certain portions of the body, and diarrhœa of a peculiar kind, do not appear until the disease is fully developed, and in numerous instances are not observed at all. The animals feed fairly well, show no rise of temperature; they are vigorous, and for several days, and even for several weeks, betray in no way the disease. It requires very careful and prolonged observation to notice that at one time or another—at any rate at rare intervals—the swine have a short cough, not an important departure from the healthy state. Very few experts, to say nothing of the owner of the pigs, would suspect swine fever because he happened to hear the animals cough at rare intervals; and there is in these mild cases, which Dr. Klein considers constitute the majority, absolutely nothing else which would be noticed; the skin remains quite free from any appearance of disease. It is true that the inguinal glands are distinctly enlarged, but unless the owner thinks of inspecting these small lymphatic glands (kernels) under the skin of the pig's groin, and knows what their proper size is in health, he has no chance of recognizing the disease in this obscure form. In the organs of these apparently healthy pigs there are found on dissection very pronounced symptoms of disease, so pronounced in fact, that it seems astonishing that during life the animals presented such slight signs of it. And these very slight signs, it may be remarked, were only noted by the observer in the cases of pigs which were under constant observation after they had purposely been infected with the disease.

Swine which are affected with swine fever in the occult form may be moved about freely, sent to market, bought and sold over and over again, distributing the infection wherever they go, without exciting the least suspicion in the minds of those who buy or sell or keep them that they are infected with a deadly and highly contagious disease, and in this way many outbreaks may occur without the origin of the infection being discovered.

In the more pronounced forms of swine fever the symptoms which are shown by the sick animal will not be very definite until the affection is fairly advanced. Dullness and diminished appetite, with hot skin and occasional shivering fits, with rise of internal temperature from the normal 103° to 105° or 106° F., are among the first signs of infection, and should always be taken as justifying a strong suspicion of the existence of swine fever, especially if the disease is known to be in the locality.

It has been remarked that pigs when suffering from swine fever in the early stage often seek to hide themselves beneath the litter on the floor of the sty; but this action is so common among swine that it would hardly be likely to attract any attention, and even when noticed it would not be looked upon as a symptom of any disease. There is, however, to the practiced eye a great difference between a healthy pig, which, in its desire for warmth, or quiet, or for some other reason, buries itself under a lot of litter, and one which performs the same act in obedience to the instinctive effort of a sick animal to get out of the way of the light and its companions, and everything under the sun, and suffer in seclusion. The healthy pig when disturbed emerges from its retreat alert and ready for action. The sick one objects to move, and if compelled, crawls from his lair, trembling and discontented.

Sometimes there are signs of partial paralysis, and the pig moves in an unsteady manner from side to side, frequently losing the use of one or both hind quarters, and dropping to one side, or dragging both hind legs as it attempts to move forward. There is no loss of sensibility with the defective motion, as the animal if touched will indicate pain by a sharp cry.

Diarrhea may occur early in the disease, after a short period of constipation; and the evacuations are generally light in color at first, becoming darker by and by, often from the mixture of effused blood. In severe cases the intestinal discharge consists almost entirely of blood, with masses of clot and loosened pieces of exuded lymph from the inflamed and ulcerated mucous membrane of the digestive canal.

A symptom which is considered to be characteristic of swine fever may sometimes be detected early in the disease. Red patches or blotches appear behind the ears, inside the arms, and under the belly. Professor Axe describes a distinct eruption in the form of red spots, from one to three lines in diameter, slightly raised above the surface, so that they may be felt by drawing the finger over the skin. The eruption is not always present, and often there are not more than half a dozen spots extending over a large surface; but in other cases they are very numerous. After two or three days the spots subside, and are followed by a second, third, or even a fourth crop, and after their final disappearance the cuticle becomes ragged and scales off.

Small water bladders (vesicles) appear on the surface of the red spots when the fever remains very high, and the fluid contained in them either escapes or becomes dried up, forming a gray crust on the surface.

Discharges of thin fluid from the eyes and nose commonly takes place early in the course of the disease, and, as it advances, the discharge becomes thick and purulent, sticking about the eyelids and openings of the nostrils. Diarrhea is more constant as the disease goes on. The voided matter is offensive and often streaked with blood; prostration occurs and is soon followed by complete collapse, and the animal dies in a state of unconsciousness or in violent convulsions. Death may take place at various periods from the commencement of the affection, sometimes after a few days' illness, while in other cases the animal may linger for two or three weeks.

In reference to the change of the color of the skin in marked cases of swine fever, it may be observed that there may be, independently of the raised spots which Professor Axe describes, a diffused redness, or regular patches of redness, or a purple tinge in different parts of the surface; but these signs are not to be expected in all or even in the majority of cases of the disease; and it is well known that redness of the skin is a symptom of some common diseases of swine. It arises, for example, from exposure to wet and cold; contact with sea-water, or spray during a voyage, will also produce it; a journey in a railway truck which has been lime washed is another cause; consumption of wash with which salt liquor has been mixed is followed by severe, often fatal, inflammation of the mucous membrane of the digestive canal and sympathetic irritation of the skin; and the symptom has also been noticed in acute indigestion which has resulted from eating freely of mangolds. Redness of the skin, therefore, either in patches or generally, can not be accepted as sufficient evidence of the existence of tho swine fever.

In certain instances the symptoms of swine fever may be so well defined during life that an opinion may be given promptly and without much risk of error, but under ordinary conditions the conclusion must be the result of the careful consideration of the history of the outbreak, or of the evidence which can be obtained by dissection of the organs of a diseased animal; and for this purpose it is better to kill a suspected pig than to take the carcass of one which has died some time before the commencement of the inquiry, and has probably undergone *post-mortem* changes which will render obscure the true morbid appearances.

Experience has placed beyond all doubt the fact that swine fever when it once obtains a hold on a herd of swine does not spare them. It is true that a proportion, and it may be a considerable one, of the older animals particularly, will escape with a mild attack, and in some of these mild cases there will be no characteristic symptoms to attract attention. In fact, it will appear that animals have escaped the disease; but the critical observer will find some signs of the disease in the majority of the swine which have been exposed to infection. Cases which are often reported of swine fever attacking two or three swine of a large herd and sparing the rest may safely be put down to the credit of any disease but swine fever. And the same thing may be said of those cases which are traced to the consumption of indigestible food, or the exposure of the animal to hardship or unsanitary conditions of existence.

Post-mortem appearances in swine fever.—Some of the changes which are affected in the organs of the body by the ravages of swine fever can only be appreciated by the pathologist; others, and those most distinctive perhaps, are patent to the ordinary observer; and on this point it is worth while to note the description which was given before the members of the Royal Agricultural Society in 1865 by Dr. William Budd, of Clifton, who was the first scientist in this country to investigate the disease, which was at that time looked upon as a new one. Dr. Budd's attention was particularly attracted to the changes which had occurred in the digestive organs, which he called a series of ulcerations of peculiar character, variously distributed over the intestinal tract, from the stomach to the rectum inclusive.

The first stage of the local affection appears to be the development (amid all the phenomena of acute inflammatory disturbance) in the substance of the mucous membrane and in the submucous tissue of an adventitious deposit (or cell growth) resembling in many of its characteristics the well-known yellow matter of human typhoid. When Dr. Budd gave this description, the outgrowth from or deposits on the mucous membrane of the stomach and intestines were seen more frequently than they have been of late years; in fact, in all well marked cases they were the prominent objects in the *post-mortem* appearances—soft, spongy masses, sometimes in very severe cases stained a dark red by the blood, which was extravasated into the intestines, varying in size from that of a large pea to a walnut, circular or oval in shape; sometimes several masses were joined together; but whether large or small, separate or confluent, the deposits, from their color in contrast with the membrane on which they were placed, were very striking objects.

A very curious appearance was sometimes seen in the intestines of swine which had partially recovered from the fever and afterwards died from exhaustion, or were killed for the purpose of examination. The mucous membrane in these cases was spotted with the small masses of deposit which had lost the soft, spongy or fungoid character, and become dense and, so to speak, leathery looking, not unlike yellow leather buttons, marked with concentric rings, or, as Dr. Budd remarks, like slices of columba root.*

The first simile will perhaps be more suggestive to the farmer than the latter, as slices of columba root are not familiar objects. The mucous membrane of the intestines, especially the large intestines, was often covered with ulcers, and the masses of deposit which have been referred to were generally found to be in connection with deep excavations, the result of the ulcerative process; but in all cases the edges of the ulcers were elevated above the surface of the mucous membrane, and presented a soft fungoid character.

In some cases the whole of the mucous membrane of the intestine was covered with a croupous or diphtheritic deposit of a dirty white color; and it was only after the deposit had been cleared away that patches of inflamed, and perhaps ulcerated, structure could be seen.

In some instances the diphtheritic deposit was so abundant as to fill the intestinal tube, and leave no canal for the passage of the feculent matter, and rupture of some part of the intestine was the natural result of this blocking up of the passage.

The next illustration is copied from a report on swine plague in the Report of the Bureau of Animal Industry, Washington, United States (for 1885), and it may serve to indicate at the same time the characteristic appearances of the disease and the identity of the swine plague (hog cholera) in the United States of America and the swine fever of this country.

Besides the appearances which have been described as occurring in the stomach and intestines, especially in the large intestine, there are nearly always observed patches of congestion in the lungs, with here and there condensation of the lung structure, which present a fleshy character quite distinct from the healthy state. There are also changes, easily distinguished by the pathologist, in the liver, kidneys, spleen, lymphatic glands, and also in the cavities of the heart. In short, it may be affirmed that the morbid changes in swine fever are so perfectly well defined that no error in opinion can occur when all the evidence which a *post mortem* inspection affords is in possession of the inquirer.

In this description we are led to infer that the writer has before him genuine hog cholera. At the same time some of the features dwelt upon, such as the croupous and diphtheritic deposits or exudates in the large intestines, are more like those found in severe epizootics of swine plague.†

There is another fact which proves that at least part of the swine fever of England is the hog cholera of the American continent. This is presented by the investigations of E. Klein. In his report upon swine

---

*These neoplasms we have also encountered chiefly in adults. They must be looked upon as results of a greater resistance on the part of the mucous membrane, such as we may expect in old animals.

†A plague like hog cholera may perhaps change its character under influences such as climate and locality, or under influences found in race, breed, etc., of animals. This may frequently account for modifications of diseases caused by the same micro-organism.

fever in the supplement to the Seventh Annual Report of the Local Government Board for 1877 Dr. Klein gives a very good account of the pathological anatomy of this disease, together with some researches on the nature of the cause and the mode of infection. Though calling the disease pneumo-enteritis, his text, as well as the statements of Professor Brown, quoted above, show that the accompanying lung disease is no more severe than in hog cholera. The microbe which he described in this report is a large spore-bearing bacillus. The methods which he employed in his bacteriological work were those of the times, not considered now as capable of giving any trustworthy information. Consequently, at a later date he regarded it as necessary to again go over the ground, and published his results in Virchow's Archiv, XCV, 408, in 1884. In this monograph he describes the inoculation disease in mice and rabbits very much as we have found it, laying particular emphasis upon the coagulation-necrosis found in the liver. His cultures contained a motile bacillus from .002 to .003 millimeter in length. We may therefore conclude that a germ closely resembling or identical with the hog cholera bacillus is found in cases of swine fever, and that hog cholera actually exists in England.

The history of the struggle with swine fever in Great Britain since 1878 shows that it has been very severe. Numerous orders have been issued, beginning with the contagious diseases act in 1878, to restrict the extension of the disease, but all appear to have been without avail, since the malady was as widespread and as virulent in 1885 as ever before. This was without doubt due to the inefficient and incomplete execution of the orders. By the order of December 17, 1878, the slaughter of swine affected with the disease was made compulsory, and compensation was to be paid out of the local rates for all swine slaughtered by order of the local authorities. The order also provided that no swine should be moved out of a pig-sty or other place without a license, and then only for slaughter. After this order had gone into effect weekly returns were received from the inspectors of local authorities, which showed that in 1879 the disease had prevailed, at some time during the year, in forty-four counties in England, six in Wales, and three in Scotland.

The continued spreading of the disease gave rise to the swine fever order of 1879. This contained provisions for the declaration of infected places and for the slaughter of healthy swine which had been herded with diseased ones, and also for regulating the exposure for sale of swine in fairs and markets. This order was only partly carried out, for out of 17,074 swine attacked 3,416 were allowed to die and 124 recovered. In 1882 the disease was more prevalent than in 1879. Slaughter of diseased and exposed animals was continued as before. Thus of 14,763 swine attacked 11,903 were slaughtered, 2,799 died and 18 recovered. An order of council was passed in July, 1882, providing for the declaration of swine fever infected areas by the privy council, and giving power to local authorities in England and Wales to prohibit or regulate the movement of swine into their districts from the district of any other local authority in Great Britain. The order also empowered local authorities in Scotland to prohibit or regulate the movement into their district of swine from the district of any other local authority in Scotland; it did not appear, however, that the local authorities availed themselves of these powers, except in very few cases.

In 1884 the area of the disease rather increased, while the number attacked did not grow larger. Further orders of council were passed this year containing provisions relating to swine fever. The most important change made referred to the provision for slaughter of diseased swine, which was objected to by some local authorities as a costly and comparatively ineffectual measure. By an order passed in May the compulsory provision was revoked, and the slaughter of diseased swine, as well as those herded with them, was left to the discretion of the local authority. Another order of council, passed in July, 1884, provided for the formation of a swine fever infected circle round an infected place in any district to which the order might be applied by the privy council at request of a local authority. The order was applied during the year to five counties and ten boroughs.

In 1885 the disease grew to dangerous proportions, threatening the entire stock of pigs in the country. The extension was due to the purchase of swine at fairs and markets, and it was therefore deemed expedient to impose restrictions on the sale of swine in all parts of England. Accordingly an order of council was passed prohibiting the exposure for sale of swine except for slaughter within three days, and then only with the license of the local authority of the district in which the sale was to be held. Provision was also made for the exhibition of swine at agricultural shows.

This so-called markets and fairs order did a great deal of good, but, as it did not prohibit private sales, the disease even increased in some counties, owing to the unrestricted sale and movement of swine.

"It was continued in operation until the end of November, and during its operation the centers of disease were so far advanced in number that it was deemed expedient to have recourse to slaughter of diseased swine, and also of those which had been in the same pig-sty or shed with diseased swine. With this view the swine fever compulsory slaughter order was passed, and came into operation on the 1st of December. The order provided that local authorities should slaughter all swine that at any time in the month of December were affected with swine fever, and all swine being or having been in contact, or in the same pig-sty or shed, with swine affected with swine fever.

"Compensation was to be paid to the extent of half the value of a diseased pig and the full value of a healthy pig, but the amount was not to exceed 40s. for a diseased and £1 for a healthy pig. During the four weeks to the end of 1885, the effects of the system of compulsory slaughter were not altogether satisfactory. In fact, fresh outbreaks of swine fever continued to occur in the same districts, and even in the same premises, owing to the way in which the order was carried into effect in regard to the slaughter of swine which had been exposed to infection. For example, if a lot of swine were in a sty which was separated a few feet from the sty in which disease existed, the animals were not slaughtered, because they could not be said to be in contact, and they were clearly not in the same sty, although it might fairly be presumed that the swine so situated were exposed to risk of infection through the atmosphere, or by the agency of persons or things by which the virus of the disease could be conveyed to a considerable distance. * * *

"Slaughter of diseased and infected swine was adopted in some districts with satisfactory promptitude. In others a number of animals were allowed to die before any action was taken. The disease, after being stamped out in some districts by the energetic action of the authorities, was reintroduced from other parts of the country where the law was administered in a perfunctory manner. In short, nothing like a serious and decided effort was made to get rid of the affection, and as a natural consequence the good effects of restrictions were partial and temporary.

"Imperfect cleansing and disinfection of premises may be reckoned among the causes which have contributed to the continuance of swine fever, notwithstanding the operations of regulations which might have been expected to produce good results. Very frequently swine are kept in places which can not be cleansed and disinfected effectually, so as to make them safe for the next lot of pigs which will be brought in as soon as the place is declared free. Old, half rotten styes with moldy floors can not be cleared of infection by any known process. The only course in such cases is to remove the infection-saturated timber and soil and submit them to the action of fire; but no power is vested in the authorities to do this necessary work, and the only expedient which they can employ is that of refusing to declare the infected premises free until the necessary alterations and improvements are completed. In this somewhat roundabout way they may attain the object—that of keeping out fresh swine to become victims to the disease, but this kind of pressure can only be applied under certain circumstances.

"Free movement of infected and of actually, although not observably, diseased swine, and their exposure in fairs and markets, has all along been a fruitful cause of the spreading of swine fever. Many inquiries have been made with the view of tracing the origin of different outbreaks, and the conclusion has generally been that they were due to introduction of swine recently purchased at public sales.

"Under the system, which has obtained for years, of slaughtering diseased swine, and allowing the apparent'y healthy to live, the extinction of the disease could not possibly be secured. And it is only in the common course of things that the malady has gradually extended all over the country from the sale and movement of diseased and infected swine which often showed no signs of disease, and would be passed as free from swine fever by any one but an expert who is familiar with every phase of the malady. * * *

"In reference to the measures which should be adopted for the extinction of swine fever in this kingdom, it is not necessary to enter into details. On three occasions in the last twenty years the stamping-out system was successfully applied to cattle plague. Exactly the same means would certainly succeed in ridding us of swine plague, and there is no ground for the expectation that any less severe measures of repression would be effective. It is, however, an essential condition of success that the action taken should be uniform and general in application. But it is impossible to suggest any means of securing the necessary uniformity while the execution of the law is in the hands of some hundreds of local authorities who entertain different views as to the necessity for attempting to get rid of the disease by legislative measures, and are not agreed as to the proper means of effecting the object."

SWEDEN AND DENMARK.

In the fall of 1887 a disease closely resembling hog cholera appeared in Sweden among swine, which rapidly spread from place to place, showing itself very fatal and causing alarm and consternation among the agricultural population. The disease invaded the territory of Denmark, where stringent measures were adopted to prevent its further spread. The following communication, received by the Department of State from our consul at Copenhagen, and kindly forwarded, may serve to illustrate the measures employed by the authorities in checking the epizootic. After giving briefly the symptoms of the disease, and the lesions caused thereby, Mr. H. B. Ryder continues as follows:

The very prompt and stringent measures taken by the Danish Government. it is to be hoped, will be the means of localizing as well as of effectually stamping out this malignant pest. For example, circulars have been sent from the home department to all the sheriffs throughout the kingdom, instructing them to make publicly known that all persons who within the last two months may have purchased hogs or young pigs from Copenhagen, or in its vicinity, should immediately give notice thereof, so that their entire herds might be submitted to veterinary inspection. Furthermore, an ordinance has been issued strictly prohibiting all transport of live hogs or young pigs from one district to the other , and that no removal of the animals shall be made from their present dwellings, excepting by special permission for the purpose of immediate slaughter; and, lastly, power has been given to all the police authorities. on any suspicion of disease, to order the immediate slaughter of the animals, and a *post-mortem* examination of the carcass to be made by the veterinary officials; and on the appearance of the disease in any locality under their jurisdiction to order the immediate slaughter of a part or the entire herd as may be deemed necessary.

It is thus to be hoped that by these measures further spread of the disease may be arrested. It is, however, much to be feared that, in addition to the losses entailed upon the kingdom in the destruction of animals in the course of the disease, the sorely tried farmers in these days of agricultural depression will also be subjected to material loss in a diminished sale to foreign countries of their swine products. An order has already been issued by the German federal council prohibiting all importation into the German Empire of swine, pork. and sausages from Sweden, Norway. and Denmark, which will be most seriously felt by the agricultural interests. as the exports of live hogs and young pigs are almost entirely directed to Germany, whilst the exports of pork and hams are mainly forwarded to Great Britain, as will be seen by the following table, namely:

Countries.	Average 1881–1885.	1885.
Of live hogs and pigs to—	*Head.*	*Head.*
Germany	276,166	192,273
Norway..	6,844	4,704
Great Britain...	2,895	128
Holland..	1,243	....
Of pork and hams to—	*Pounds.*	*Pounds.*
Great Britain...	12,550,000	20,240,000
Germany..	2,930,000	7,180,000
Sweden	3,970,000	3,030,000
Norway..	1,200,000	700,000

From the foregoing figures full evidence is afforded that whilst the exports of live stock have met with considerable decline in the latter years a great increase has taken place in the exports of swine products, due to the large number of slaughtering and salting establishments which have been erected in this country for the development of the pork, bacon, and ham trade with England; and thus the loss to the agricultural interests, it is to be hoped, will not be quite so severely felt as would have been the case in former years under similar unfortunate conditions, and it is scarcely to be feared that England will likewise place obstacles in the way of the free imports of the products into her ports; for inasmuch as the imports of swine into Great Britain from this country have for some time only been admitted in slaughtered condition, and setting apart the facts that swine in mature stage for slaughter are far less exposed to this disease than young pigs, there will be found at the same time, under the close inspection which has been introduced throughout the Kingdom, and the energetic steps taken by the authorities in all cases of suspicion, an almost certain probability that no pork from a diseased animal can possibly be exported. The sale of swine products for home consumption plays likewise a very important part; and it is here again to be feared that restricted sales will be sensibly felt until the temporary scare in partaking of swine flesh has had time to subside.

In order that the energetic measures taken by the Government for the stamping out of the plague may be crowned with full success it will be necessary that the agriculturists should give at the same time a loyal support to the issued instructions and work hand in hand with the authorities. He who may delay in reporting or may attempt to conceal any disease or suspicion of disease that may show itself at his place will simply be committing a crime against the class to which he belongs, and honesty in this as well as in other cases will be found the best policy; for whereas he who reports the breaking out of disease amongst his stock will receive two-thirds of the value of the slaughtered diseased animals, and full compensation for the slaughtered sound ones, the dishonest party will incur not only risk of confiscation of the diseased meat offered by him for sale, but will also render himself liable to heavy fines. The closing of Germany to the importation of these products undoubtedly can not fail to entail severe loss upon the agricultural classes; but if success can only attend the stringent measures adopted for preventing the further spread of the disease it must be hoped the prohibition will be of short duration, and that agricultural interests will soon recover from the blow. But should the devastating plague, on the other hand, spread over the whole Kingdom it will be nothing short of a national calamity, the destructive effects of which will long be felt. as will easily be understood from the following table of the number of hogs and young pigs to be found in the Kingdom under census of 1881, viz:

	Hogs.	Young pigs.
	Head.	*Head.*
In the islands...........	285, 317	392, 884
In Jutland...............	242, 100	334, 707
Total	527, 417	727, 591

Great responsibility will thus rest not only upon the veterinary and police authorities but also upon the agriculturists in devoting all possible energy in their mutual exertions to prevent the further spread of this dreaded evil.

The disease is supposed to have been introduced into Sweden by diseased boars, imported from England for breeding purposes. Thence it was carried to Denmark,* in which country it first appeared on the dumping grounds near Copenhagen, on which numbers of swill-fed pigs were kept.

Chiefly young pigs up to the age of four months were attacked, the period of incubation lasting from five to twenty days. The infected animals refused food and were at first constipated. Later on diarrhea set in, characterized by the discharge of yellow putrid masses, frequently mixed with blood. The temperature often rose to 105°–107.5° F. The animals were indifferent to surroundings. Tail and head drooping. Conjunctiva reddened, frequently glued together with dried-up mucus. Respiration in many cases quickened and labored. Occasionally a mucopurulent discharge from the nose. Not infrequently reddening in patches was observed on the ears, snout, abdomen, about the anus, and inner side of thighs. The animals became very weak; posterior part of body swayed in moving about. They concealed themselves in the bedding, and finally were unable to rise. Death followed insensibility and convulsions.

A characteristic sign of this plague were diphtheritic changes on the apex, sides, and under surface of the tongue, as well as on the mucous membrane of the cheeks, hard and soft palate, and the tonsils. On these parts grayish white or yellowish opaque patches appeared, which were sharply defined and were converted later into ulcers by removal of the slough.

In one herd the teats of several sows were affected with dark gray, sloughing sores, with inflammation of the udders. These were infected

* Schütz: *Die Schweinepest in Dänemark.* *Arch. f. wiss. u. prakt. Thierheilkunde*, XIV, 1888. p. 376.

from the diphtheritic sores in the mouths of the sucking pigs. In Denmark the disease was first recognized in this way. The acute disease lasted from five to eight days, but sometimes death occurred sooner than this.

The disease appeared in Denmark in September. In December the plague took on a more chronic character and became less infectious. The infected animals frequently showed no indications of disease, only they were smaller and thinner than others of the same age. There was occasionally coughing and diarrhea. Some recovered, others perished by a gradual wasting away. The *post-mortem* appearances were very characteristic. The large intestine was attacked in every animal, and in acute cases the small intestine and stomach likewise were reddened and swollen, and the surface in part covered with a thin layer of a grayish white or grayish yellow soft mass, which consisted of fibrin. This same layer was very thick in the large intestine, easily lifted away *in toto* in some places; in others the attachment was firmer (diphtheritic). In other acute cases there was simply reddening and swelling of the mucosa of stomach and small intestine, and diphtheritic changes in the large intestine, the fibrinous exudate being absent. Moreover, the follicles, Peyer's patches, and mesenteric glands were always tumefied.

The seat of the diphtheritic process was the whole large intestine, more especially the cæcum. The follicles and Peyer's patches were nearly always affected. The ulcers appeared when the slough had come away. In the place of the follicles button like sloughs were formed, which gradually invaded the whole thickness of the wall, spread laterally, and ran together into larger patches. The wall thus converted into a cheesy mass was frequently one-fifth to two-fifths inch thick, on the surface irregular, colored yellow, brown, or green. Hemorrhage, due to the ulceration, was observed in one case.

In many animals the lungs were healthy. In some a muco-purulent catarrh of the bronchi was present, which caused atelectasis in one or more places with young and weak animals. Usually the ventral and anterior lobes were affected. In the diseased lobes homogeneous, cheesy masses appeared later, sometimes as large as walnuts. These masses led subsequently to inflammation and adhesion of the pleura to chest-wall, pericardium, etc. The spleen was not changed as a rule. In a few cases only it was somewhat enlarged, soft, dark red.

When we compare these lesions with those found in our country, we observe the absence of hemorrhagic lesions and enlargement of the spleen, and more marked exudative and diphtheritic changes in the large intestine. In numerous sections of ulcerative changes we have not observed any relation between these and the follicles. The lung lesions correspond closely. Whether they are due to the disease or not must be left undecided. We have frequently seen caseous changes in the lungs of animals free from infection, and they are, perhaps, due to collapse, broncho-pneumonia, and subsequent interference of the circulation, rather than to the direct action of bacteria.

The specific bacteria which are the cause of the swine disease are described briefly by Selander,* and according to his description they closely resemble hog cholera bacteria in form, motility, growth in gelatine, and appearance in tissue. Their growth on potato is said to resemble that of the bacilli of typhoid fever in man, and thus to differ from that of hog cholera bacilli. Their effect on the lower animals correspond also, although the descriptions are too brief for careful comparison. There is no mention of the coagulation necrosis found constantly in the liver of rabbits inoculated with American hog cholera.

In the beginning of the present year (1888), Dr. John Lundgren, professor of veterinary medicine in the University of Stockholm, was sent by the Swedish Government to study swine diseases in this country. He spent several weeks in the laboratory of the Bureau studying the bacilli of hog cholera. A culture of the swine pest bacilli from the Swedish epizootic was at that time subjected to a careful examination.

In gelatine, the swine pest germs grow very much like hog cholera bacilli. On the surface of the gelatine the growth is very thin, translucent, of a pearly luster, and spreads more rapidly than the hog cholera growth. On *agar-agar* the growth is more abundant and more rapid. Beef infusion, with or without peptone, is converted into a very turbid liquid within twenty-four hours at 95° F., while hog cholera cultures are barely opalescent at that time, and remain so. Two mice were inoculated from an *agar-agar* culture of the Swedish germ under the skin of the back. Both were slightly ill next day. On the second day one was found dead. The cultures from it remained sterile. It probably died from some other cause. The second mouse remained well. On a rabbit the effect was equally negative. No rabbit survives inoculation with hog cholera bacteria.

The effect of both germs on pigs was next tried. Two Erlenmeyer flasks, containing each about 300 cubic centimeters (two-fifths of a pint) of sterile bouillon, were inoculated, one with the American, the other with the Swedish germ, and placed in the thermostat at 95° F. On the following day both flasks were clouded ; the Swedish culture was covered by an iridescent, very thin membrane. A comparative microscopic examination showed the Swedish bacteria to be nearly twice as large as the American ; their movement was far less active than that of the latter.

On the same day two pigs, starved for about twenty four hours, were fed with these cultures by drenching, i. e., the liquid was poured into the mouth, so that none was lost. The pig fed with the Swedish culture showed no signs of disease at any time after. The other pig on the fourth day had a very liquid diarrhea, and was found dead the next morning. On examination the spleen was found gorged with blood, but only slightly enlarged. Mesenteric glands enlarged and reddened.

---

* *Centralblatt f. Bakteriologie, etc.*, 1888—i, 362.

Stomach and ileum intensely inflamed (enteritis); grayish masses (diphtheritic) attached in patches. The ileum was invaginated and projected for 2½ inches into the cæcum; mucosa of this portion necrosed; walls infiltrated, thickened, and ecchymosed. In cæcum the mucosa was covered by a very thin slough. In the colon the membrane was deeply reddened, covered by a catarrhal exudate, and dotted with numerous very minute ulcers. Heart and lungs normal. Roll cultures in gelatine as well as liquid cultures from the spleen contained only hog cholera bacteria. The invagination was very likely the result of the violent inflammation.

These comparative experiments show that the two germs, though very much alike in appearance, were quite different with reference to their pathogenic effect. Professor Lundgren was inclined to the opinion that he had taken the wrong culture on leaving his native country. It may also not be improbable that this was the true germ attenuated on the way hither. As no communication has been received from him since his visit here, the question must remain an open one.

FRANCE.

During the summer of 1887 a disease was introduced into the vicinity or Marseilles by swine from Africa, which developed into an epizoo'ic of a very fatal character. It caused great losses in the south of France, and at the time scientific men were sent from Paris by the Government to investigate the cause and suggest a remedy, if possible. According to Rietsch, Jobert, and Martinand * the disease is chiefly restricted to the intestinal tract, lasting from ten to twelve days after the first symptoms have appeared. Occasionally it may last but three or four days, or be prolonged to several weeks, but it is quite invariably fatal. Sometimes there is diarrhea, sometimes constipation; the fever is not constant, the cough very rarely heard. The hind limbs are weak, the walk tottering. Appetite often persists to the end. The skin may become reddened in spots, especially on the limbs and ears. Pigs over a year old are much less susceptible.

At the autopsy the lungs, liver, kidneys, and spleen are usually found unchanged, and the disease limited to the digestive tract. The stomach and the small intestine near the valve are ulcerated. Ulcers are present in the large intestine on the valve, in the cæcum and colon. They may measure 3 to 4 inches in diameter. In animals affected with a chronic form of the disease, there may be ulcers on the inferior surface of the tongue and on the inner aspect of the lips. The internal organs are free from bacteria. But from the contents of the intestine and the ulcers a motile bacillus was obtained. Mice are killed in ten days after subcutaneous inoculation with cultures of this organism. Of ten adult mice fed with cultures two died in fifteen days. Of ten young mice all

Compt rend. Acad. Sciences, January 28, 1888.

died when fed, the first in thirty hours, the others in fifteen to twenty-three days. Rabbits are but slightly susceptible. A young guinea-pig, fed with cultures, died twenty-two days later. The intestine showed characteristic ulcers.

Dr. Rietsch very kindly sent to the Bureau a culture of the germ which he found. It was compared with the American hog cholera germ and the following characters determined:

The motile bacilli of the same form as hog cholera bacilli, but larger, grow far more abundantly and rapidly in beef infusion. A thin membrane and a copious deposit are produced in a few days, and the liquid becomes very turbid. On gelatine the colonies do not differ appreciably from those of hog cholera. The surface colonies spread in thin irides-cent patches from 2 to 5 millimeters in diameter. In tube cultures the isolated deep colonies grow to about one-third of a millimeter in diameter. On the surface the growth is rapid and spreads over the greater part of that which is available. The patch produced is whitish, uniform in thickness, very irregular in outline, inclosing round spaces of uncovered gelatine. In the bottom of the tube a few air-bubbles appear. On po-tato the growth forms a glistening, pale-yellow patch at the ordinary temperature.

They thus resemble the Swedish germs very closely, differing by so much from hog cholera bacilli. A rabbit and two mice inoculated with the Marseilles germ remained well. Equally so a pig fed with 400 cubic centimeters (four-fifths of a pint) of a beef infusion peptone culture.

So far as our examination of the Swedish and Marseilles cultures have gone, they have shown them identical both as regards their posi-tive and negative characters. They differ from hog cholera bacteria enough to constitute at least a variety. But the investigations of French savants of this Marseilles epizootic differ somewhat as to the cause.

Cornil and Chantemesse* described a disease discovered among swine in the vicinity of Paris, which they consider identical with the German *Schweineseuche* and our swine plague. Subsequent experiments† to determine the biological properties of the bacteria causing the disease show that they are not dealing with the true swine plague germ (cer-tainly not as we have observed it in this country), but with one resem-bling more nearly hog cholera. Their researches concerning vaccina-tion are reported to have been successful on rabbits and guinea-pigs, but since that time nothing has been published concerning experiments on pigs.

While Rietsch and Jobert‡ come to the conclusion that the Mar-seilles disease resembles hog cholera closely, Cornil and Chantemesse regard it identical with swine plague, although the germ they describe

---

* Compt. rend. Acad. Sciences, 1887, CV, p. 1281.
† Compt. rend. Acad. Sciences, 1888, CVI, p. 612.
‡ l. c., p. 1096.

is not identical with the swine plague germ as studied in Germany and in the United States.*

It is interesting in this connection to trace the march of infection in the south of France as reported to the French Academy by Fouquet as an excellent illustration of the ways by which infectious diseases may be scattered broadcast over a country:

" The disease did not, as was supposed, appear in Marseilles toward the end of June, but in the month of April, and I have been able to locate three entirely distinct centers of the outbreak due to the same cause—the introduction of African swine. These three centers are : The village of Caillols, midway between Aubagne and Marseilles; the village of St. Marthe, 6^{km}. northeast of Marseilles, and the herds of the Mediterranean distilleries.

(1) From the 10th to the 15th of April, a breeder of Caillols received a drove of black swine from the province of Oran (Algiers). From the first week some cases of pneumonia‡ showed themselves among the last animals; the disease gained rapidly, causing many deaths. The survivors were sold on the 4th of May following.

On the 8th of June the same piggeries were restocked, partly with African and partly with Russian swine. Towards the end of the month there were cases of pneumonia. The Russian swine resisted less than the others. August 16 the piggery was again emptied. Finally during September the third attempt was made, exclusively with African swine. This also proved a failure. The survivors were sold in October.

During this time the disease reached the neighboring piggeries stocked with a mixed Marseilles breed. The breeders of Caillols, alarmed by the ravages of an epizootic the nature and cause of which they did not know, decided to sell out at any price. The neighboring localities, St. Marcel, St. Loup, San Joan du Désert, etc., were successively infected. Infection was spread by the sales and exchanges of sick and suspected animals, by means of transportation (carts often used in common by several establishments soiled by the dejections of the sick and afterwards used to transport healthy animals and their feed), and also by the lateral canal of Huveaune, which receives at certain points running water coming from the grounds on which the piggeries are located.

At the beginning of September all the valley of Huveaune, from Aubagne as far as Marseilles, was infected. Diseased pigs from this region we meet again in the market of Aubagne, at the fair of September 21, and which became later one of the most active agents in spreading the disease in the departments.

(2) Toward the middle of the month of August the disease appeared in a piggery in St. Marthe, stocked exclusively with African swine. These animals came directly from Oran without coming in contact with any other of their species. Several days later one of the largest breeders in that vicinity, who for three months had not brought a single pig to his establishment, and whose piggery was at least 600 feet away from the preceding one, sustained a considerable loss, especially among the pigs of 130 to 175 pounds.

(3) Finally, on the 25th of June, sick pigs came from Oran to the piggeries of the Mediterranean distilleries. There were very soon a number of victims of pneumonia, not only in the distilleries but also in the neighborhood, where there were from 4,000 to 5,000 in a comparatively small territory. A great many of the sick died ; the others were quickly sent to different cities to be delivered to the butchers. I have traced the history of six sows which were sold from the midst of infection to Estaque; thence they passed through the commune Rove and arrived in August at Gignac, where they introduced the disease. By an odd coincidence some sick pigs from the same locality were taken to the fair at Aubagne and bought by a breeder of Gignac.

The fair at Aubagne, on September 21, marked the most important phase in the progress of contagious pneumonia. During the first fortnight in October there was a veritable explosion of the disease, which, up to this time, had been scarcely known.

The importation of the disease by animals bought at the fair of Aubagne can be traced with the greatest precision in the suburbs south and north of Marseilles, also as far as Gardanne, in the communes of Septemes, Vitrolles, Pennes, etc., to Gignac, as mentioned above, even into the neighboring departments, which continued with the others to receive consignments of Marseilles swine. It is also necessary to mention Puget, Ville et Grasse, among the localities infected.

In the beginning of December 153 swine were shipped from Marseilles to Nice; nearly all died in a few days. From that time cases appeared among the native pigs.

On December 22 another lot of 133 pigs were shipped; 33 were destined for Nice and 100 for Italy. These last were sold on the 24th, in the market of Vintimille; nearly all died very soon.

For several years Marseilles annually exported to Spain, and especially to Barcelona, a great number of pigs. Contagious pneumonia had been causing losses there for several months, even, it is said, at Majorka, in the Balearic Islands. The Spanish breeders, believing the outbreak of the disease with them was attributable to the importation of French pork, obtained from the authorities the permission to impose a quarantine of six days, at Port Bouc, on swine from Marseilles, to begin on the 1st of the following February. This measure has not been enforced up to the present day.

From what has preceded I believe I can conclude that the epizootic of contagious pneumonia which has raged during the year 1887 in the interior of France, and which at this time continues its ravages there, is of African origin. It has been introduced by Algerian swine which came from the province of Oran. This disease has made 20,000 victims in several months in the province of Bouches-du-Rhone.

Pigs, and especially those from three to nine months old, are oftenest attacked; larger pigs appear less susceptible. The Marseilles breed, the English (Yorkshire and Berkshire), and the Russian swine are more susceptible than the African swine.

Two months ago about 50 pigs two to three months old, coming from Cazeres and Le Fousseret, in the arrondissement of Muret, were used to stock a farm in Gignac. These pigs, placed in the pens which had contained sick ones, and which had only been very imperfectly disinfected, remained in good health, while more than a hundred cases of contagious pneumonia appeared around them in the same piggery. Afterwards more than 2,000 Gascon swine were imported by the single commune of Gignac. Up to the present time the disease has not re-appeared. Are we here confronted by a new example of natural immunity comparable to that noticed long ago by Chauveau in Algerian sheep in regard to anthrax?"

Taking into consideration what we know now of these epizootics and enzootics of swine diseases in foreign countries, we are forced to the conclusion that the disease in Sweden, Denmark, and France is closely related to if not identical with hog cholera as it is found in our own country, while the minor differences in the disease as manifested in these epizootics, and in the germs producing it, can be explained on the principle of the variability of disease germs, a principle in favor of which much evidence has already been obtained in recent bacteriological researches.

DESCRIPTION OF PLATES.

PLATE I:

Cæcum of a pig affected with hog cholera laid open to show the ulcers of the mucous membrane. The ileo-cæcal valve is near the center of the page, and the small intestine, with the cut end tied, is above. The smallest ulcers have sloughs of a uniformly yellowish color; the larger ones have zones of different color, while the largest are brown or blackish. It will be seen from the plate that the slough or new growth in most ulcers projects slightly like a small flat button.

PLATE II:

Cæcum of a pig fed with viscera from a case of hog cholera, slit open to show the mucous membrane thickened and quite uniformly necrosed, with isolated deeper ulcerations. The ileo-cæcal valve is very much thickened, its mucous membrane ecchymosed and ulcerated. The lymphatic glands of the meso-colon and those in the angle formed by the entrance of the ileum into the cæcum are purplish, with cortex engorged with extravasated blood. They illustrate the condition of the lymphatics of both thorax and abdomen in the acute hemorrhagic form of the disease.

PLATE III:

Cæcum of a pig inoculated with blood from a case of hog cholera. The entire mucous membrane is necrosed. In the upper portion of the figure about the valve there are groups of minute pigment spots. The valve is open to show the intact mucosa of the ileum. This figure also serves to illustrate the condition of the mucous membrane of cæcum and colon in pigs fed with cultures of hog cholera bacilli.

PLATE IV:

Cæcum of a pig, showing round and elongated cavities on and around the ileo-cæcal valve. These are ulcers from which the slough dropped away during removal. The base of the ulcers is formed by the muscular tissue of the intestinal wall. The ulcers appear as if the mucosa had been punched out with a sharp instrument. The edge of the ulcers in the upper portion of the colon with the slough still adherent. The lungs in this animal were normal, if we except a few collapsed areas from one-half to three-fourths inch across.

PLATE V:

Ulcers in the lower portion of the small intestine (ileum); not very frequent. Note the large, deep ones with puckered, ragged border, and the very small ones with a thin superficial slough, stained with bile.

PLATE VI:

Kidney from pig No. 7 (see p. 41), showing the hemorrhagic condition of cortex and engorged glomeruli. In this animal the disease was marked by the hemorrhagic condition of other organs.

191

PLATE VII:

Right lung from a case of hog cholera. The lung tissue is nowhere infiltrated or hepatized, but studded with hemorrhagic spots which are found in the parenchyma as well as under the pleura. This animal, placed in an infected pen, died on the twenty-second day of exposure. There were hemorrhages in the subcutaneous connective tissue; lymphatics in general hemorrhagic. Petechiæ in stomach and large intestine. In the cæcum and on the valve a few ulcers with hemorrhagic base.

PLATE VIII:

Heart with subepicardial hemorrhages especially numerous on left auricle.

PLATE IX:

FIG. 1. Collapse of various groups of lobules in the principal lobe of a pig's lung. Frequently found in young pigs which have died of hog cholera, in those affected with lung worms, and in a small percentage of those slaughtered during health.

FIG. 2. Broncho-pneumonia affecting animals under various conditions and not infrequently found in animals which have succumbed to hog cholera. (See p. 54). The air cells and smallest air tubes are shown distended with a yellowish material which is of a dry, caseous consistency, and may be teased out in the form of minute branching cylinders.

PLATE X:

FIG. 1. From a cover-glass preparation of the spleen of a pig which died of hog cholera. Stained in an aqueous solution of methyl violet and mounted in balsam. Outlined with a camera lucida, using a $\frac{1}{18}$ homog. immersion (Zeiss), ocular 2 (x 800): a, distorted red blood corpuscle; b, bacilli in pairs.

FIG. 2. Section from the enlarged and congested spleen of pig No. 94 (see Second Annual Report Bureau of Animal Industry, p. 193), which died very suddenly with extensive hemorrhagic lesions of various organs. Spleen hardened in alcohol two hours after death; stained in aniline water methyl violet and decolorized in 1 per cent. acetic acid; mounted in xylol balsam; outlined as in Fig. 1. Note the large group of bacilli occupying the capillary meshes of the spleen pulp, all extra-cellular.

PLATE XI:

Illustrative of the growth of hog-cholera bacilli on gelatine.

FIG. 1. Various colonies from gelatine plates: a, embedded in the gelatine layer (deep colony), usually spherical or egg-shaped; b, flattened colony growing on the surface of the gelatine layer; c, c', c'', growing between the layer of gelatine and the glass plate as very thin films; the knobs on c', c'' represent the part of the growth in the gelatine layer.

FIG. 2 represents both deep (a) and surface colonies (b) growing in a gelatine layer coating the inner surface of a test tube. A tube of gelatine was inoculated by stirring up in it, while liquid, a minute fragment of spleen tissue from a case of hog cholera (No. 464, p. 51), transferring a drop from this to a second tube, and then rolling out the latter in ice water. Culture seven days old. The zones on the surface colonies are very likely due to changes of temperature in the laboratory, alternately retarding and augmenting the growth. This mode of growth is by no means common.

FIGS. 3, 4, 5 represent the growth of hog cholera bacilli in tubes of nutrient gelatine, showing the isolated colonies in the depth of the gelatine and the confluent growth on the surface. Fig. 3 represents a culture three days old (epizootic near Washington City, 1885), inoculated from the spleen of a pig. Fig 4, fourteen days old, inoculated from a culture which was prepared directly from the spleen of a pig in Illinois, 1886. Fig. 5, inoculated from the blood of a rabbit which died from inoculation with bacilli from Illinois culture; ten days old. The difference in growth observed may be due to a greater or less alkalinity of the culture medium or to a slight physiological difference in the bacilli themselves.

PLATE XII:

Fig. 1. *Agar agar* culture of hog cholera bacilli in thermostat, four days old.

Fig. 2. Potato culture under a bell glass, twelve days old.

Fig. 3. Potato culture in a test tube containing an abundance of moisture. The plug was made impervious with sterile paraffine. The growth is whitish, very glistening. The difference in color between Fig. 2 and Fig. 3 is caused by the different conditions of moisture.

PLATE XIII:

Fig. 1. Coagulation necrosis in the liver of a rabbit which died six days after inoculation. The small, yellowish spots are groups of acini completely necrosed. The larger patch to the left shows the necrosis in the form of a network.

Fig. 2. Liver of a rabbit inoculated with one thirty-second cubic centimeter (by dilution) of a beef infusion peptone culture one day old. Dead on the seventh day. The coagulation necrosis on the left is more diffuse. On the right are cysts of *coccidium oviforme*.

Fig. 3. Enlarged spleen from the same rabbit (natural size). Weight of rabbit, 1¾ pounds. The spleen of this rabbit was but moderately enlarged as compared with the spleens of most rabbits which succumb to hog cholera.

Fig. 4. Spleen from a healthy rabbit weighing 2¾ pounds (i. e., one and a half times the weight of the diseased rabbit).

PLATES XIV, XV, XVI:

Photomicrographs of hog cholera bacilli which developed in different media. All were made at a uniform magnification of 1000 diameters, with the Zeiss apochromatic objective of 3ᵐᵐ and 1.30 numerical aperature, using projection ocular No. 4 and Abbe condenser with largest diaphragm. Orthochromatic plates and picric acid screen.

PLATE XIV:

Fig. 1. Coverglass preparation from spleen of rabbit inoculated with hog cholera bacilli. Stained two to three minutes in aqueous solution of fuchsin. Mounted in xylol balsam. × 1000.

Fig. 2. Coverglass preparation from bouillon-peptone culture five days old. Stained in aniline-water-fuchsin for five minutes, and decolorized in one per cent. acetic acid for a few seconds. Mounted in xylol balsam. × 1000.

PLATE XV:

Fig. 1. Coverglass preparation from bouillon-peptone culture one day old. Stained same as Fig. 1 of Plate XIV. Mounted in xylol balsam. × 1000.

Fig. 2. Coverglass preparation from gelatine culture two days old. Stained same as Fig. 2, Plate XIV. × 1000.

PLATE XVI:

Fig. 1. Coverglass preparation from colony of hog cholera bacilli taken from an Esmarch tube, made directly from spleen of pig, fifteen days old. Stained same as Fig. 1, Plate XV. × 1000.

Fig. 2. Coverglass preparation of hog cholera bacilli from *agar* culture fifteen days old. Stained same as Fig. 1, Plate XIV. × 1000.

15612 H C——13

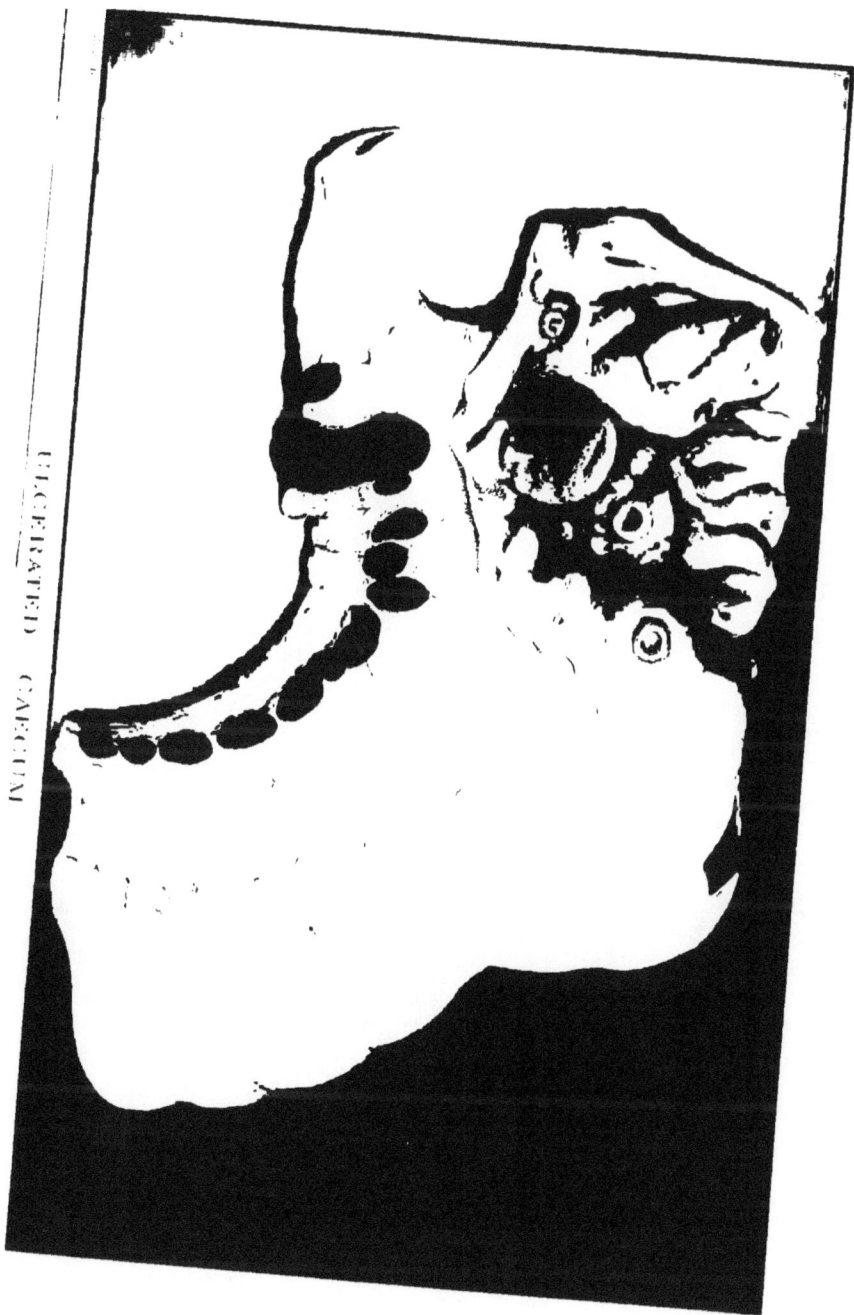

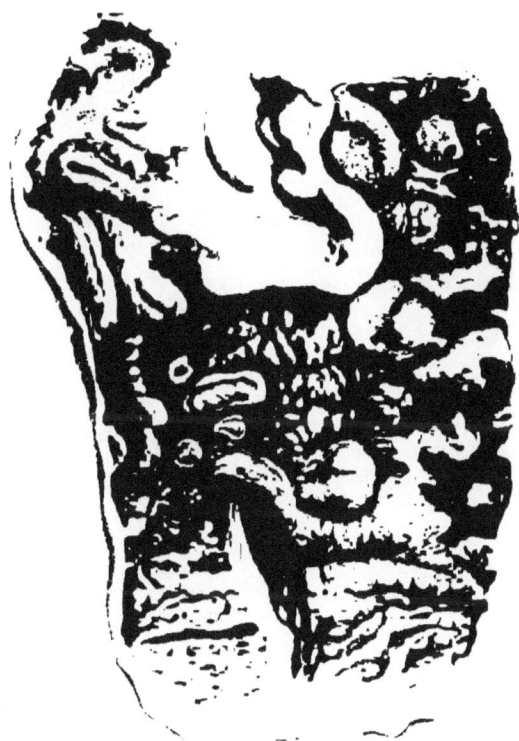

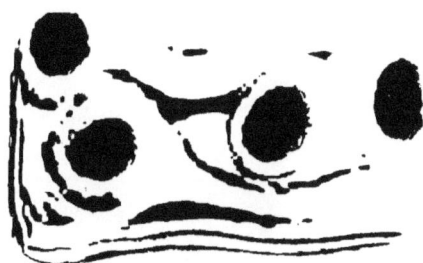

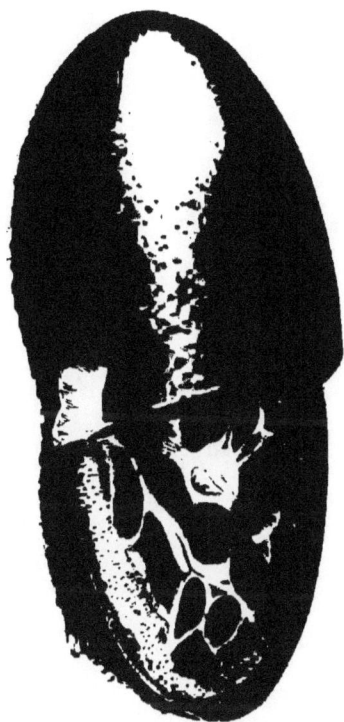

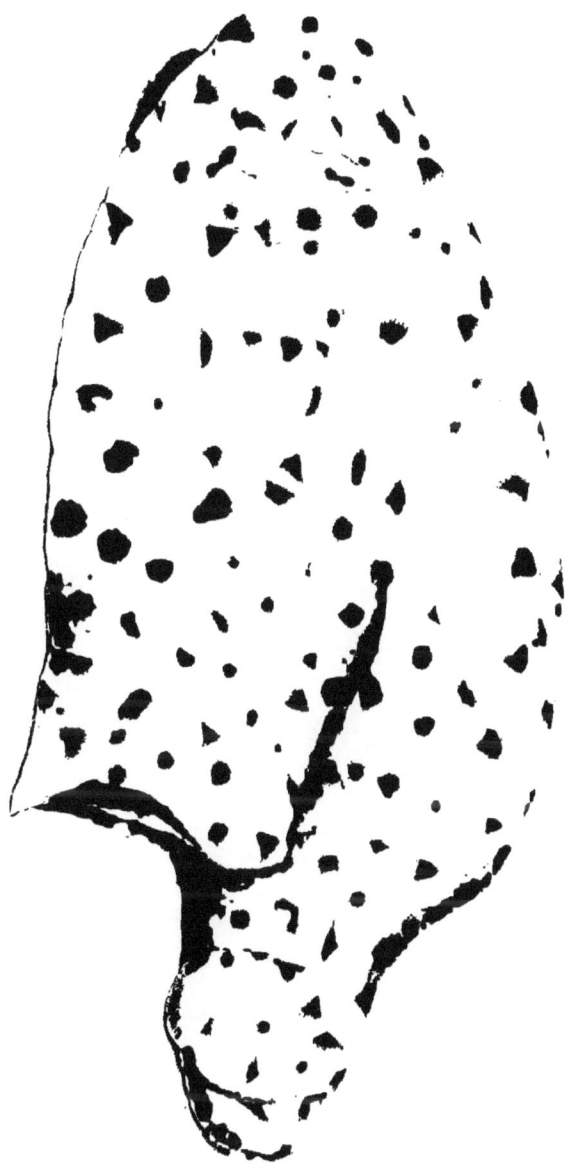

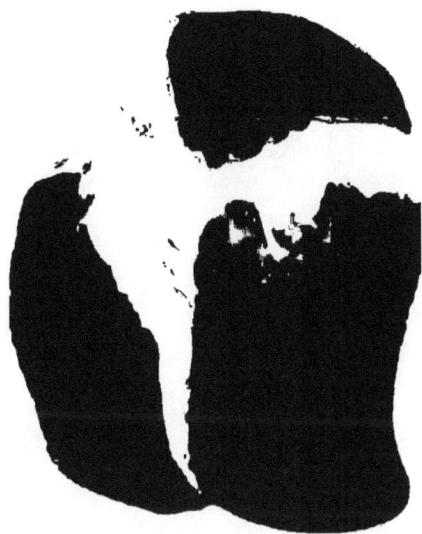

Fig. 1

Fig. 2

Fig.1
b

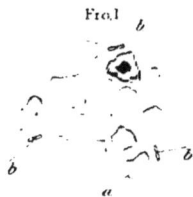

b

a

Fig 2

HOG CHOLERA BACILLI.

Plate XI.

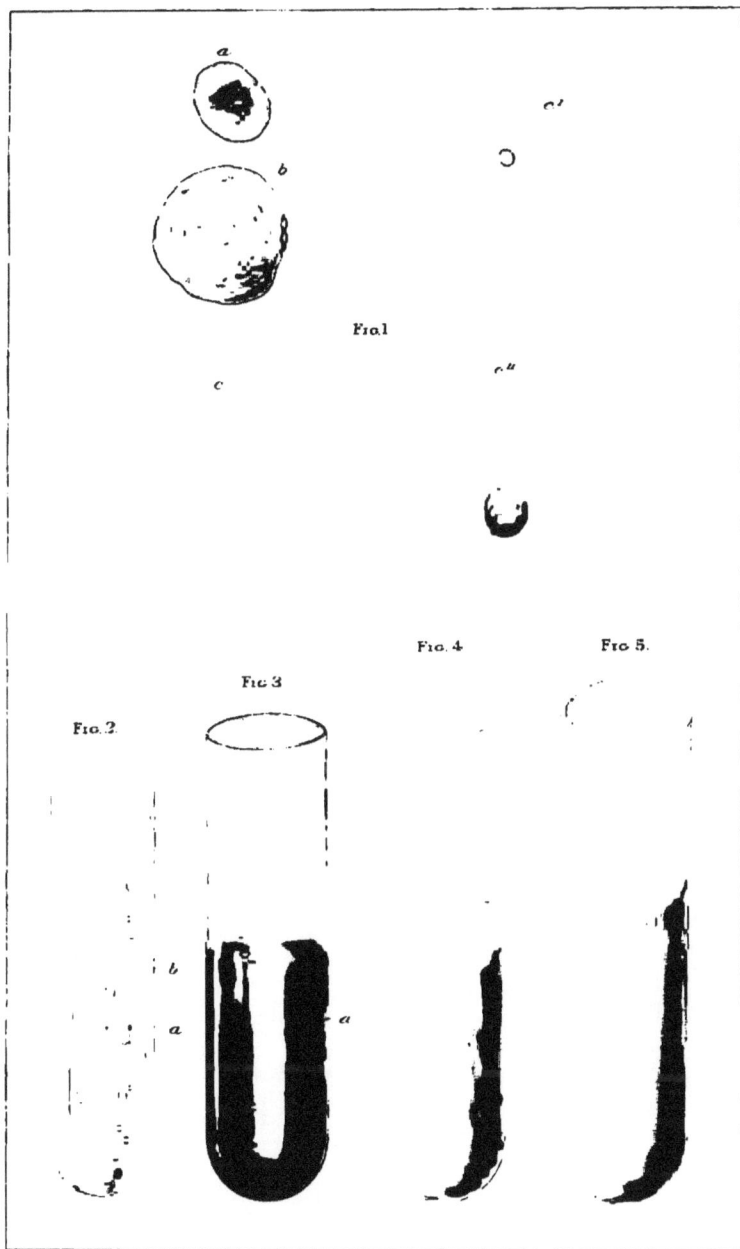

a

c'

b

c''

Fig. 1

c

Fig. 4   Fig. 5.

Fig 3

Fig. 2

b

a   a

Marx fecit

CULTIVATION OF HOG CHOLERA BACILLI.

Fig. 2.

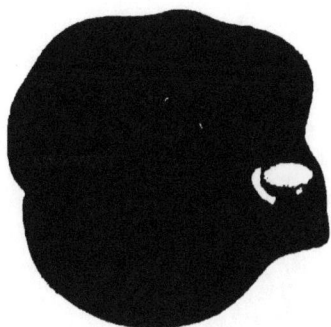

Fig. 1

Fig. 2

PHOTOMICROGRAPHS OF HOG CHOLERA BACILLUS.

FIG. 1. x 1000.

COVERGLASS PREPARATION FROM SPLEEN OF INOCULATED RABBIT.

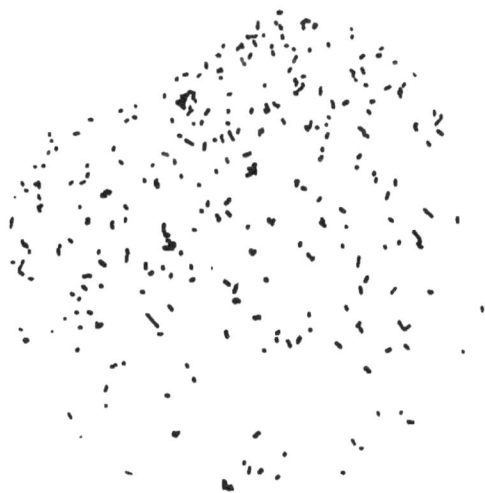

FIG. 2. x 1000.

COVERGLASS PREPARATION FROM LIQUID CULTURE FIVE DAYS OLD.

PHOTOMICROGRAPHS OF HOG CHOLERA BACILLUS.

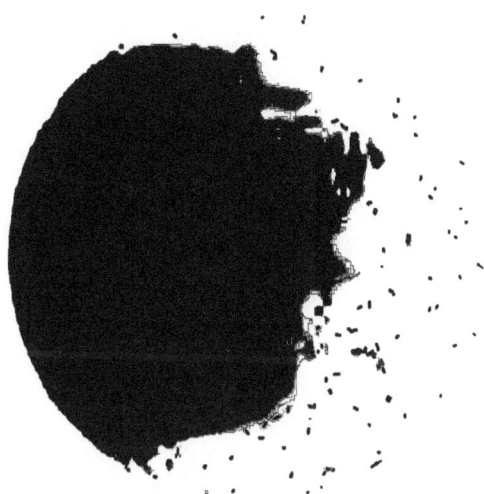

FIG. 1. x 1000.

COVERGLASS PREPARATION FROM LIQUID CULTURE ONE DAY OLD.

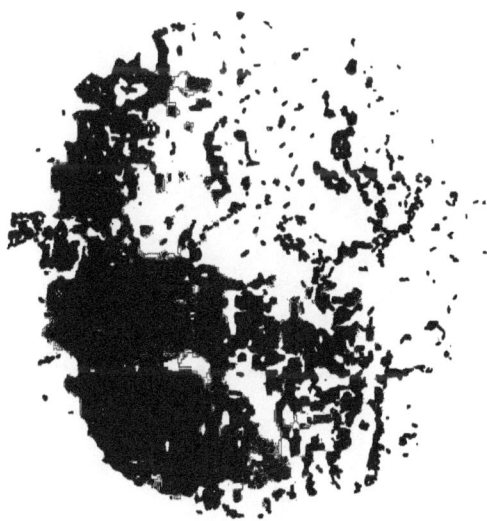

FIG. 2. x 1000.

COVERGLASS PREPARATION FROM GELATINE CULTURE TWO DAYS OLD.

FIG. 1. x 1000.
COVERGLASS PREPARATION FROM ROLL CULTURE (ESMARCH TUBE)
FIFTEEN DAYS OLD.

FIG. 2. x 1000.
COVERGLASS PREPARATION FROM AGAR CULTURE FIFTEEN DAYS OLD.

INDEX.

○